SpringerBriefs in Geography

SpringerBriefs in Geography presents concise summaries of cutting-edge research and practical applications across the fields of physical, environmental and human geography. It publishes compact refereed monographs under the editorial supervision of an international advisory board with the aim to publish 8 to 12 weeks after acceptance. Volumes are compact, 50 to 125 pages, with a clear focus. The series covers a range of content from professional to academic such as: timely reports of state-of-the art analytical techniques, bridges between new research results, snapshots of hot and/or emerging topics, elaborated thesis, literature reviews, and in-depth case studies.

The scope of the series spans the entire field of geography, with a view to significantly advance research. The character of the series is international and multidisciplinary and will include research areas such as: GIS/cartography, remote sensing, geographical education, geospatial analysis, techniques and modeling, landscape/regional and urban planning, economic geography, housing and the built environment, and quantitative geography. Volumes in this series may analyze past, present and/or future trends, as well as their determinants and consequences. Both solicited and unsolicited manuscripts are considered for publication in this series.

SpringerBriefs in Geography will be of interest to a wide range of individuals with interests in physical, environmental and human geography as well as for researchers from allied disciplines.

More information about this series at http://www.springer.com/series/10050

Arthur Lehner

The Operational Use of Remote Sensing in Municipalities

A Global Expert-based Study

Springer

Arthur Lehner
Department of Geoinformatics
University of Salzburg
Salzburg, Austria

ISSN 2211-4165 ISSN 2211-4173 (electronic)
SpringerBriefs in Geography
ISBN 978-3-030-23799-8 ISBN 978-3-030-23800-1 (eBook)
https://doi.org/10.1007/978-3-030-23800-1

This Springer imprint is published by the registered company Springer Nature Switzerland AG
The registered company address is: Gewerbestrasse 11, 6330 Cham, Switzerland

Acknowledgements

I want to acknowledge all experts that were so kind to provide their expertise and time for the conducted interviews.

Introduction

By now, the use of remote sensing has made its way into all conceivable areas of application. Among others, remote sensing techniques are operationally used in agriculture (Klisch and Atzberger 2016), agronomy (Baruth et al. 2008; Sakaiya and Inque 2013), forestry (San-Miguel-Ayanz and Ravail 2005), mining (Martini et al. 2004), navy (Olsen et al. 1995), and marine science (Saitoh et al. 2011).

When it comes to cities and urban agglomerations, urban remote sensing encompasses all activities regarding the acquisition of information about the urban environment and physical objects within urban areas through recording, measuring, and interpreting electromagnetic radiation derived from contactless sensors (Lehner and Steinnocher 2016). According to Kadhim et al. 2016, recent advances in satellite remote sensing offer opportunities to cost-effectively monitor urban change and its impact on the complex urban socio-technical systems and enable stakeholders to make informed decisions to reduce negative impacts on the environment, such as urban sprawl, the destruction of habitats, or the reduction of recreation areas. In the context of knowledge about remote sensing technology in general, Specter 1988 conducted a survey by means of a questionnaire, asking experts around the world about the transfer of (US) remote sensing technology (e.g., Landsat satellite imagery) from the USA to other countries and possible obstacles. Additionally, improvements in small-scale technology have enabled unmanned aerial vehicles (UAVs) as an alternative remote sensing platform, offering a distinctive combination of very-high-resolution data capture at a significantly lower survey cost (Wallace et al. 2012).

In a nutshell, the literature often states remote sensing (RS) data are nowadays available globally and at any required resolution (Netzband et al. 2007; Bhatta 2010; Esch et al. 2010; Yang 2011). However, do city administrations really make use of this availability and benefit from it? Does the researcher's perception really coincide with the prosaic reality that urban planners and other administrative staff have to face in their day-to-day business?

This study answers the question whether RS as such and RS-derived products are operationally used within city administrations worldwide for tasks in urban planning in general (Herold et al. 2002; Netzband et al. 2007; Bhatta 2010) and energy planning in particular (Tooke and Coops 2013). Additionally, I analyze which RS-derived products are used in participating administrations and for which purpose. Furthermore, I explicate how the competences of the responsible departments, divisions, or sections are defined in this context.

For this purpose, I carry out expert interviews with persons responsible for RS data and products for the respective cities. The information provided in these interviews and hence drawn conclusions are presented in a condensed form. According to Bogner and Menz 2009, expert interviews are by no means simply just "information gathering meetings" used primarily for collecting facts and knowledge. Rather, the authors point out that expert interviews are not only a popular way of gathering information but also a totally legitimate method for some forms of research.

The difficulty in this study originates from the circumstance that there are thousands of cities, yet none of them are spatially or morphologically identical (Fig. 1) nor do they share a common base in organizing their RS data (if present) or their geodata, respectively. They do share an administrative background in whatsoever form. Nevertheless, the existence of an administrative unit or governmental order does not necessarily indicate the (distribution of) responsibilities and tasks therein. In general, the administrative organizations and the definition of their competences are mostly intangible for people outside these organizations. Further, as later presented in the result section, the distribution of competences is not as crisp as it may appear from the outside when it comes to, for example, the acquisition of RS data, the production of maps, and the provision and distribution of available RS data. Recently, there has been an ongoing shift (Agrawal et al. 2014) in governmental and citizens' awareness toward the open government, open data, or government transparency initiatives in general. By means of these tools and policies,

Fig. 1 The structural and spatial heterogeneity of cities, including cultural and political differences, complicates the application of one overall organizational solution for data management and administrative problems in general

governments and citizen initiatives address these diverging perspectives and aim to develop an interface between citizens, governmental institutions, and their respective work.

Independent of the publicly available background information, this study taps into expert knowledge, which provides researchers with facts concerning the question they are investigating (Bogner and Menz 2009). For this purpose, a questionnaire has been set up. The challenge remains to review the obtained information, to derive an overall picture, and to meet the claim of answering the research question of whether city administrations use RS for urban planning and energy planning, respectively.

Contents

Chapter 1
The Operational Use of Remote Sensing in Municipalities: A Global Expert-Based Study

1.1 Study Areas and Background Information

The existing diversity between municipalities, in terms of facilities, staff and technology, depending on their country or region, is undisputed. This circumstance warrants an investigation into the operational use of RS, not only for one city of one country, region or continent, but also for different cities, located in different climate zones, and, especially, for cities with disparate economic backgrounds (Fig. 1.1).

1.1.1 Study Areas

I contacted 73 cities and organizations, respectively, located in 41 countries on six different continents (cf. Appendix Table A1) to meet the above-mentioned requirement of incorporating case studies with varying climatic, social, and economic backgrounds. My selection of cities was further strongly dependent on the existence of an appropriate point of contact. This means that contact information of any form (email-address, general contact form, fax, phone) had to be available online. However, in cases where contact information was not available online, I was able to contact some RS experts based on personal recommendation. A comprehensive list of all contacted cities and institutions is given in the appendix, Table A1.

1.1.2 Background Information

Additional information used for the analysis includes information folders and general principles of particular departments, divisions or institutions, provided

A. Lehner, *The Operational Use of Remote Sensing in Municipalities*, SpringerBriefs in Geography, https://doi.org/10.1007/978-3-030-23800-1_1

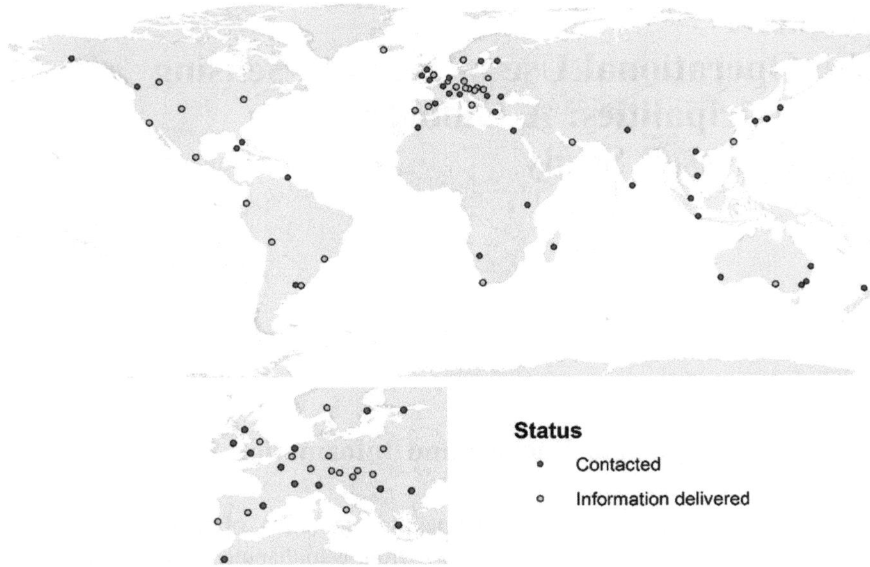

Fig. 1.1 World map of contacted cities and institutions

through contractors or available online (list of websites: cf. Appendix Table A2). The website quality of the contacted municipalities or respective institutes, who are responsible for the management and distribution of RS data and resulting products, strongly deviates; the same holds true for the quantity of the provided information through these websites. Ancillary data and background information presented in the tabulated summaries of the particular cities and institutions contacted are derived as follows:

- National GDP in US-Dollar (World Bank Group 2017; International Monetary Fund 2018),
- National population (national statistics, WorldPopulationReview 2016),
- Municipal population (national statistics, WorldPopulationReview 2016),
- Administrative area in km^2 (national statistics, municipal websites).

1.2 Expert Interviews

The objective of this study is to gather and access the comprehensive knowledge of RS experts located in a specific city or region on a global scale in a bid to answer the posed research question. I access the knowledge of local experts by means of RS expert interviews.

For conducting these interviews, I have set up a questionnaire that consists of twelve questions. The structured questionnaire used for the expert interviews serves as a base to convey the information to a qualitative and quantitative evaluation. The questionnaire contains open questions (e.g., "How is data management and the distribution of remote sensing data and resulting products organized?") and closed questions (e.g., "Are remote sensing data provided free of charge or may additional costs incur?"). The closed questions serve for comparability among the studied cities and institutions and facilitate the identification of differences among the administrative entities. The open questions allow the experts to provide as much information as desired and to deliver insights as much as possible. In this regard, this research primarily uses a descriptive and qualitative approach (DiCicco-Bloom and Crabtree 2006; Berg and Lune 2017) to answer the research question whether RS and resulting products are operationally used in municipalities or not.

The questions of the questionnaire remain the same for all contacted cities and institutions. For almost each country, the questionnaire was adapted and translated into the respective official language. With the aid of background information, the questionnaire was adapted in the sense of mentioning local departments' or divisions' names without changing the general structure of the questions or their meaning. At the beginning, there were uncertainties regarding the number of questions and their ambiguity. However, after conducting the first interviews it became evident that the number and length of the questions were adequately chosen, and the content was clear. The twelve (groups of) questions are as follows (Table 1.1).

Table 1.1 Questions for the RS expert interviews

1	Does department XY play the main role in the acquisition of aerial imagery or remote sensing data respectively (satellite images, aerial images, and terrestrial data)? → Can other departments or divisions (e.g. department of AB, division 123) take action and acquire/purchase remote sensing data on their own, or does department XY function as the interface and represent the first and exclusive point of contact?
2	Which remote sensing platforms or data are used respectively? (Aerial imagery, 3D laser scan data, radar data, high-resolution satellite imagery, terrestrial laser scan data?)
3	How is data management and the distribution of remote sensing data and resulting products (orthophoto, DEM, DSM) organized? (Intranet platform, web form, personal requests?)
4	Are the Orthophoto and the DEM/DSM the only products derived from remote sensing data that are used within the city administration? What other products are provided by department XY and which ones are produced using remote sensing data?
5	Which municipal departments or divisions usually use the service of department XY and use remote sensing data or resulting products respectively?
6	For which tasks or activities do these departments/divisions normally use remote sensing data or resulting products respectively? (Nature protection, tree cadaster, building regulations?)
7	Does documentation exist about which departments use remote sensing data and for which tasks?

(continued)

Table 1.1 (continued)

8	Do requests for remote sensing data exclusively come from communal areas (within public authorities) or also from private entities, such as companies, extra-faculty (non-university) research institutes, university institutes, or individuals?
9	Are remote sensing data provided free of charge or may additional costs incur?
10	Is there a web portal from which remote sensing data or resulting products can be acquired?
11	Do you think that the importance of remote sensing within public authorities is increasing, decreasing, or remaining steady?
12	Do you know if services like Google Street View, Google Maps, Bing Maps, etc. are used within your department or other departments of the city administration?

1.3 Investigated Countries, Cities and Institutions

Thirty-one of the 73 contacted cities and institutions responded with sufficient answers, hence covering 22 studied countries; i.e. a return rate of 42%. Generally, large and capital cities, respectively, were targeted for contact and expert knowledge inquiries. Based on these experts' answers, it is hypothesized that the answers of the respective cities are not only exemplary but also representative; i.e. based on the presented way of using and managing RS data, the general (operational) use of RS data for urban planning and energy planning purposes, respectively, can be deduced for the whole particular country. Usually, the operational use of RS data is not limited to the classic departments, such as the ones responsible for the urban environment (protection), urban planning or building authority; as such, RS data is widely used in all departments.

1.4 Insights into the Global Municipal Use of RS

RS has long arrived in urban administrations. The study shows that RS fulfills far more than just "simple" monitoring functions; in many municipalities, countless mapping products and other products resulting from RS data are now produced and used by the departments (e.g., planning departments, environmental departments, transportation departments etc.) themselves. Municipalities have developed different strategies regarding the question of which institution is responsible for the acquisition of RS data. Acquisition is conducted either in-house with particular equipment or by private companies in free competition. Due to their size, some cities are relinquishing the acquisition of RS data and merely use the RS data provided by superordinate regional governments. The decisions and ways about how collected or purchased data are made available or provided to other departments are very different. In many municipalities, data is stored centrally and, with a few exceptions, are equally accessible to all departments. While some administrations have an infrastructure for interdisciplinary data exchange, others have no ordinance requiring departments to share raw data or RS data derived products,

Table 1.2 Examples of currently in use RS data and resulting products

Governmental/Municipal entity	RS data used	Product/Support of
Graz	Thermal infrared aerial imagery	Urban climate analysis
Vienna	Aerial imagery and LiDAR	Noise cadaster
Warsaw	LiDAR	Heritage protection
County of Los Angeles	Aerial imagery and LiDAR	Solar radiation model
State of São Paulo	Satellite imagery	Watershed protection area mapping
State of South Australia	Aerial imagery (includes hyperspectral data) and satellite imagery	Benthic mapping

such as vegetation maps, visibility analyzes, or evacuation analyzes, with other departments. In such cases personal contacts foster the exchange of RS data and resulting products. Therefore, it can be stated that communication between departments plays an important role. In order to increase the flow of communication between individual departments and to improve existing products, some cities have evaluation cycles within the administration in which questions are asked regarding existing maps, analyses and the satisfaction with such products and other services. These evaluation forms are sent out to the departments and evaluated in specific cycles. Measures such as additions to map materials, additional aerial acquisitions or the introduction of new RS data (UAV data, Oblique recordings) may follow.

The single-case character as already present in the acquisition and the provision of RS data can also be found in the choice and use of RS derived products. In order to give a brief foretaste of the diverse applications to be found in the following sections, exemplary products are presented that are currently used in municipalities worldwide (Table 1.2).

The premise of this study is to investigate the use of RS worldwide and to present the results of the analysis factually, without prejudices and judgements. I would like to leave it to the unbiased reader to read, analyze, and then possibly evaluate which approaches and procedures in respect to the use of RS can promote interdisciplinary cooperation and transparent planning processes in the future. The basis for this analysis and evaluation are the provided insights into the work of urban planning with RS presented on the following pages.

1.4.1 Europe

1.4.1.1 Austria

In Austria, three cities were investigated regarding the use of RS data and resulting products within their respective administration: Graz, Salzburg, and Vienna.

Since 1989, RS has been established in the city administration of Graz. The City Surveying Office (German: Stadtvermessungsamt) is the central office for the procurement of RS data and derived products (Table 1.3). It acts as a service provider for all other city departments.

For the city of Graz, the magistrates, the airport operating company (German: Flughafen Graz Betriebs GmbH), the holding company Graz Energy and other municipal service companies (collectively referred to as the "Haus Graz") use the City Survey Office (CSO) or its unit of photogrammetry (German: Referat für Photogrammetrie), respectively, as their first point of contact. The tasks and uses are, therefore, diversified and the departments and municipal service companies are well connected. Documents and information are collected by the CSO or via its responsible unit, and are subsequently processed, provided, and then distributed via web services.

The use of RS data is not handled authoritatively within the Graz city administration. This means that the procurement of new data, in addition to the data in stock, depends on the project and the task. Thereby, the CSO acts as an interface and/or directly as a data collector. RS data and resulting products used and provided by the CSO are as follows:

- Aerial imagery: orthophoto, oblique imagery, DTM and DSM by means of stereo matching; last image acquisition in the year 2015,
- LiDAR data: DTM and DSM, hybrid DTM,
- Thermal infrared aerial imagery,
- Very high-resolution satellite imagery,
- Terrestrial laser scan data.

Particular departments, such as the building and construction authority or the urban planning department, have great interest in oblique imagery. RS experts of

Table 1.3 Overview of the city of Graz

Overview	
Continent	Europe
Country	Austria
National GDP per capita in USD	52,558
National population	8,750,000
City population	290,000
Institution/City	Graz
Department responsible for remote sensing data and resulting products	City Surveying Office
Administrative Area in km^2	128
Number of used remote sensing platforms	5
WebGis Service	Yes
Services like Google Street View, Google Maps or Bing Maps	Yes (in the past)
Considering the importance of RS within public authorities as	Increasing

the city of Graz state the flights for image acquisition are motivated by two reasons, which influence the time of acquisition:

1. Summer solstice: coverage of the city-wide tree population (most vital point in time); recognition of terrain transformations: ideal point in time: foliage- and snow free.
2. Interval acquisition flight: for the sake of updating the data stock, monitoring, time series analysis etc.

Terrestrial laser scan data are not regularly acquired. They are used for purposes like the project-related documentation of buildings in the scope of UNESCO world heritage monitoring activities. Very high-resolution satellite imagery has been project-related only. Optical satellite imagery represents an alternative to aerial photography when no aerial photographs exist or the existing aerial photographs do not meet the desired criteria. Sometimes, high-resolution satellite data is combined with thermal scanner survey data (e.g., for urban climate analysis). An exemplary project was a thermal scanner flight in the winter in order to determine the heat loss via roof surfaces. In addition to this, optical satellite imagery is used where it provides additional information to existing data. Radar data such as SAR data have not been used.

Data management and the distribution of RS data is internally organized by means of a GIS and related databases. There are about 900–1000 information layers, such as layers with thematic references such as the building map (German: Baugrundkarte), the zoning plan, climate-related maps, etc. Generally, distribution is organized by means of an intranet and internet, where over 100 web services are available. In addition to this, an open government data platform (OGD) allows stakeholders to access RS data and resulting maps. Furthermore, there are direct analogue (printed maps or orthophotos) and digital deliveries. Certain products are only accessible to certain departments (civil protection), depending on the respective content. In addition to the existing datasets, new products are also generated, or analysis tasks performed on demand or on personal request.

RS data and the resulting products, and the provided services, are as follows: aerial imagery, true orthophotos, DTM, DSM, impervious surface maps, land cover analysis products, land cover mapping, and photogrammetrically acquired data. Tasks RS data are used for are as follows (Table 1.4).

Almost all units and departments of the city of Graz use the service of the CSO. Amongst others, the following departments of the city of Graz use the services of the CSO and, thus, RS data and the resulting products, respectively (Table 1.5).

Requests for RS data come from communal areas (within public authorities) and also from private entities, such as companies (e.g. landscape planning, architects, forest scientists), extra-faculty (non-university) research institutes (e.g. curatorship of road safety uses orthophotos), university institutes, etc.

There is no specific documentation regarding which city departments/divisions use RS data nor which tasks or activities they are used for. In a wider context, documentation is done by means of project reports and other technical reports.

Table 1.4 Tasks and activities RS data are used for

Tasks and activities
Flood protection (calculation of discharges)
Green area planning
Urban climate analysis
Control of compliance with building regulations
Urban- and spatial planning
Control of compliance with environmental regulations

Table 1.5 Departments and divisions that use RS data and resulting products

Departments and divisions
Building and Construction Authority
Department for Environment
Security Management Unit
City Planning Department
Traffic Planning Department
Cultural Office
Fire Department

In general, two-thirds of all requests have a direct reference to an urban benefit or a public entity connection, respectively. About one-third of the requests come from private entities such as companies or extra-faculty research institutes.

Data management and the distribution of RS data is organized by means of a commercially operated GIS. The costs are determined in the frame of tariff pro-visions and have to be determined by the municipal council. The respective costs are calculated based on the number of elements requested and according to the extent of the requested datasets. There is no internal cost allocation which means there are no costs for data transferred between departments internally. However, there is so-called value information for internal data de livery. This allows project costs to be calculated in a more realistic, consistent and credible way. The RS experts of the city of Graz state that the political will to open government data is present, which is also reflected in the free provision of individual data and records. They explicate that the political will defines if and to which extent (all) data should be freely available in principle, or whether only limited access to certain data shall be provided. Currently, the latter is the case, so that both freely available data exist in addition to those available at certain rates. The web portals of the city of Graz are http://www.geoportal.graz.at, and http://data.graz.gv.at/daten/geographie-und-planung.

RS experts of the city of Graz consider the importance of RS within public authorities as permanently increasing, due to the spreading awareness about the possibility of using RS data.

Services like Google Street View, Google Maps or Bing Maps have been used within the municipal organization. According to the RS experts, potential problems are caused by copyright issues. For example, the Bing Maps Service "Birds Eye

Table 1.6 Overview of the city of Salzburg

Overview	
Continent	Europe
Country	Austria
National GDP per capita in USD	52,558
National population	8,750,000
City population	153,000
Institution/City	Salzburg
Department responsible for remote sensing data and resulting products	Department 6/03 for Land Surveying and Geoinformation
Administrative area in km^2	66
Number of used remote sensing platforms	4
WebGis Service	Yes
Services like Google Street View, Google Maps or Bing Maps	Yes
Considering the importance of RS within public authorities as	Increasing

View" is not suitable for surveys or similar tasks. However, the provision of internal solutions (e.g., oblique imagery) has certainly been influenced by using services as such Bing Maps or Google Street View in the past.

Within the city of Salzburg (Table 1.6), the Department 6/03 for Land Surveying and Geoinformation (German: Magistratsabteilung 6/03 Vermessung und Geoinformation) is the central body for the acquisition and provision of geospatial (reference) data such as aerial photos (orthophotos), data from Airborne laser scanning projects or photogrammetric data. The Department 6/03 for Land Surveying and Geoinformation (MA 6/03) also acts as an external service provider for geospatial reference data within the city administration. However, the actual spatial planning agendas are carried out by the Department of Urban Planning and Transport. RS data and resulting products that are used and provided by the MA 6/03 are as follows:

- Terrestrial survey data: modern electronic tachometers for in-house data acquisition, e.g. recording of the site characteristics for the digital city map,
- GPS surveying devices: point determination based on satellite-based positioning systems (GPS and GLONASS) with Real Time Kinematics (RTK) for in-house acquisition of border points position and further survey-related operations,
- Aerial imagery: 4–5 years' time interval, generation of digital orthophotos: RGB, CIR, pan-sharpening, ground resolution at least 10 cm; in the future aerial imagery with a ground resolution of 5 cm shall be acquired. Intercommunal collaboration for image flights with the federal state of Salzburg and its respective department (SAGIS); photogrammetric evaluations for the generation of roof lines, ground marks, terrain information, etc., with a corresponding horizontal accuracy of at least 10 cm and a vertical accuracy of at least 15 cm,

- LiDAR: collaboration with the federal state of Salzburg, generation of classified point clouds, e.g. for the derivation of DSMs and DTMs.

The RS experts of the city of Salzburg state that satellite imagery has not been used within the department. The reason being that the satellite imagery does not meet the requirements regarding quality and accuracy in most cases within urban areas. However, in individual cases, other departments have acquired satellite imagery for supra-local planning projects, e.g. the department MA 5/03 of urban planning and traffic (German: MA 5/03 – Amt für Stadtplanung und Verkehr).

Data management and the distribution of RS data is internally organized by means of a GIS and related databases. The following datasets can be visualized through the respective internal web-GIS service: digital city map, cadastre, aerial imagery (orthophotos), underground infrastructure cadastre, and other department-specific GIS datasets. Most of these data and layers can only be visualized and are not directly acquirable. Special requirements regarding RS data and derived products are provided separately for in-house and external customers. A 3D city model (level of detail 'LoD' 1.5) was derived from LiDAR acquisitions of the years 2006 and 2010. This dataset is not managed by the MA 6/03 but by the Department of Urban Planning and Transport. However, the 3D city model project is currently not being further pursued by this department. In the year 2016, in the frame of a collaborative project of the city of Salzburg and the federal state of Salzburg, aerial LiDAR data were acquired. These data were used for the generation of a 3D model (level of detail 2.3) for a specific test site. According to the RS experts of the city of Salzburg, there are plans in the future to establish a comprehensive 3D model that covers the whole urban area of the city of Salzburg.

In the context of data distribution, it is worth mentioning that certain datasets are publicly available and provided free of charge on Austria's open government data platform: www.data.gv.at, and in the city of Salzburg online map: https://maps.stadt-salzburg.at. There are future plans to provide certain aerial imagery through these platforms. Further, there are plans to provide a DTM and a DSM dataset (derived from LiDAR data acquired in 2016, each 5 m spatial resolution) free of charge in the future via the city map application.

RS experts of the city of Salzburg state that, in principle, geodata of the MA 6/03 are used for all spatially related questions and tasks, which represent more than 80% of the tasks in the municipal administration. Therefore, almost all municipal departments use the data provided by the MA 6/03 to varying degrees. Amongst others, the following departments of the city of Salzburg use the services of the MA 6/03 and, thus, RS data and resulting products, respectively (Table 1.7).

The departments presented in Table 1.7 use RS data and derived products for, e.g., construction methods and proceedings, building site inspections and proceedings, generation of cadasters and maps, such as land use plan, zoning maps, etc., generation of a tree cadaster, and resource planning (Municipal Fire Brigade). Other use cases are the generation of land use maps, zoning maps, and tasks in line and pipe management. Besides this, there is no documentation regarding which city

Table 1.7 A selection of departments and divisions that use RS data and resulting products

Departments and divisions
Dpt. of Spatial Planning and Building Authority
Dpt. of Building
Chief Executive Office: Information and Communication Technology
Chief Executive Office: Economy, Investments and Land
Population and Registry Office
Municipal Fire Brigade
Municipal Archive and Statistics
Municipal Garden Office

departments/divisions use RS data, nor for which tasks or activities they are used. The remote sensing experts of the city of Salzburg state new projects are often initiated with the provision of new datasets.

Requests come from municipal authorities, from private companies (which are project-related, working on behalf of the municipality), and, in individual cases, from the field of "research and teaching" (university, universities of applied sciences etc.).

In the MA 6/03, geodata are generally issued based on a tariff sheet and corresponding third-party user rights (external parties). For special requirements, the work required for data preparation and provision (e.g., for research and teaching) is charged.

RS experts of the city of Salzburg consider the importance of RS within public authorities as increasing. They believe that geodata is generally becoming increasingly important in municipal work processes.

Services like Google Maps or Bing Maps are used to some extent by the city departments for spatial planning purposes and civil engineering.

In the city of Vienna (Table 1.8), the Municipal Department 41 of Land Surveying (German: Magistratsabteilung 41 Stadtvermessung) plays the main role in the acquisition of RS data; the Municipal Department 41 of Land Surveying (MA 41) is the first interface and point of contact for all other city departments and units requiring RS data and geodata in general. Depending on request, the MA 41 has an advisory or executive function. In exceptional cases, which are usually project-related, other municipal departments can also order remote sensing data and derived products, for example, in cooperation with university research centers. Moreover, in most cases, the other municipal departments are supported in these data acquisition and processing tasks. The RS data and resulting products used within the city administration of Vienna are as follows:

- Aerial imagery: image acquisition every year; in between these regular acquisition dates, project-related image acquisition flights may occur,
- LiDAR: the first flight with comprehensive areal coverage took place between 2006 and 2007, the second acquisition covering the whole administrative area was in 2015, project-related image acquisition flights covering smaller areas also occurred in between,

- Terrestrial laser scan data: project-related image acquisition; the Municipal Department 31 Vienna Water contracted the acquisition of such data for the survey of tunnels, used for project-related archaeological site inspections,
- Satellite imagery is hardly used because the spatial resolution of such data does not meet the requirements of the MA 41,
- Acquisition of city-wide street view imagery is planned,
- Thermal infrared aerial imagery: project-related image acquisition in the year 2012 for urban climate analysis (Jobst 2013).

Since 2015, an open government data platform is available, to access the data of the MA41 in various formats. Additionally, the MA 41 hosts a particular geodata viewer (Stadtvermessung Wien—MA 41 2016), from which RS data (e.g. tiled orthophotos) can directly be downloaded, and provides WMS and WFS services. The RS data and resulting products and the provided services are as follows: DTM (derived from LiDAR data and stereo matching of aerial imagery), DSM (derived from terrestrial data, photogrammetric data, optical RS data and LiDAR data), orthophoto, the Vienna multi-purpose map (German: Mehrzweckkarte), 3D city model in different levels of detail, a solar potential map, solar potential map for facades, and many others.

The Austrian Federal Office of Metrology and Surveying (BEV) is responsible for the land surveying of the whole of Austria. In this context, the BEV acquires aerial imagery every 3 years with 20 cm resolution. Due to the fact that the BEV acquires images every 3 years, the MA 41 aims to acquire aerial imagery for two seasons: foliage free vegetation (for the Vienna multi-purpose map) and foliate vegetation (orthophotos for visualization purposes). The RS data is usually acquired

Table 1.8 Overview of the city of Vienna

Overview	
Continent	Europe
Country	Austria
National GDP per capita in USD	52,558
National population	8,750,000
City population	1,770,000
Institution/City	Vienna
Department responsible for remote sensing data and resulting products	Municipal Department 41 of Land Surveying
Administrative area in km^2	415
Number of used remote sensing platforms	6
WebGis Service	Yes
Services like Google Street View, Google Maps or Bing Maps	Yes (in the past)
Considering the importance of RS within public authorities as	Increasing

by external providers (aerial imagery, LiDAR). The DTM, the DSM, the orthophotos, the Vienna multi-purpose map, and the 3D city model are produced by the MA 41. According to the RS experts of the city of Vienna, the most important products provided to the departments are the orthophoto and the Vienna multi-purpose map.

Almost all city departments use the services of the MA 41, but mostly those which use GIS. Planning departments, in particular, use RS data and resulting products. The following municipal departments (MA) of the city of Vienna use the services of the MA 41 and, thus, RS data and resulting products, respectively (Table 1.9).

The RS experts of the city of Vienna state the use cases for RS data and resulting products within the mentioned (Table 1.10) municipal departments are manifold. A small selection is given as follows.

Table 1.9 Departments and divisions that use RS data and resulting products

Departments and divisions
MA 8 Municipal and Provincial Archives of Vienna
MA 14 Automated Data Processing, Information and Communications Technology
MA 18 Urban Development and Planning
MA 19 Architecture and Urban Design
MA 20 Energy Planning
MA 21 District Planning and Land Use
MA 22 Environmental Protection
MA 23 Economic Affairs, Labour and Statistics
MA 29 Bridge Construction and Foundation Engineering
MA 31 Vienna Water Works
MA 33 Public Lighting
MA 37 Building Inspection
MA 41 Surveyors
MA 42 Parks and Gardens
MA 45 Water Management
MA 46 Traffic Management and Organization
MA 48 Waste Management, Street Cleaning and Vehicle Fleet
MA 49 Forestry Office and Urban Agriculture
MA 53 Press and Information Services
MA 64 Legal Affairs: Construction, Energy, Railways, Traffic and Aviation
MA 65 Legal Affairs: Transport and Traffic
MA 67 Supervision of On-Street Parking
MA 69 Real Estate Management
MA 70 Ambulance Service
City Planning Bureau

Table 1.10 Tasks and activities RS data are used for

Tasks and activities	
MA 22 Environmental Protection	Noise cadastre
	Vegetation monitoring
	Green roof potential cadastre
MA 37 Building Inspection	Compliance with building formalities
MA 42 Parks and Gardens	Tree cadastre
City Planning Bureau	UNESCO studies such as line-of-sight analysis

There is documentation regarding which city departments/divisions use RS data and which tasks or activities they are used for. This documentation is done by means of internal customer surveys.

Requests for RS data and resulting products mostly come from internal communal areas. Furthermore, there are requests from citizens for aerial imagery or orthophotos, respectively. The citizens often use these RS data as evidence to prove the existence of buildings over a certain period (building regulations: (Landtag 2001): BO §71a and b). There are also requests for GIS datasets, 3D models, etc. from private entities, such as companies and universities. In principle, there are two ways to acquire or use RS data:

1. By means of the customer service of the MA 41, including a processing service with possible processing fees.
2. By means of the open government data (OGD) portal (www.data.gv.at).

In the past, internal services were also charged, but nowadays there is no inter-municipal invoicing. For project partners, e.g. for universities and extra-faculty research institutes, RS data and services are free of charge under the condition that the results are of interest to the city of Vienna. Details of such collaborations are regulated in cooperation agreements. According to the RS experts of the city of Vienna, many data are publicly available and free of charge. For example, orthophotos for the whole city of Vienna are available free of charge for everyone (via OGD download or OGD services).

RS experts of the city of Vienna consider the importance of RS within public authorities as increasing. They justify this stance with the growing public perception of RS data and its potential use cases. In addition to this, the documentation of the national body responsible for providing safe and cost-effective management of Austria's airspace (Austro Control) shows that the number of measurement flights is growing. Services like Google Maps or Bing Maps have been used to some extent by some departments of the city. For the RS experts of the MA 41, these data and services are problematic due to the uncertainties regarding the copyright and the missing or insufficient documentation.

1.4.1.2 Belgium

In the city of Brussels (Table 1.11), the Brussels Regional Informatics Centre (BRIC) is the ICT partner within the Brussels-Capital Region entrusted with any assignment of computer, telematics and cartographic development of and assistance to:

- regional, community and local institutions,
- the offices of the Ministers and State Secretaries of the Government of the Brussels-Capital Region,
- public-interest organizations in the Brussels-Capital Region (CIRB 2018).

Thus, the Brussels Regional Informatics Centre (French: Centre d'Informatique pour la Région Bruxelloise – CIRB) plays the main role in the acquisition of RS data and resulting products in Brussels. The BRIC is officially assigned all mapping tasks by the government. In this context, the BRIC defines products and services according to the needs of users. The BRIC also defines the most appropriate methods for collecting data. The two main methods currently used for collection are large-scale aerial photography and terrestrial topographic surveys. Aerial imagery is widely used by various authorities of the Brussels Region, both at the level of municipalities and organizations that depend on the Region (Brussels Environment, Town Planning and Spatial Planning). RS data and resulting products used and provided by the BRIC are as follows:

Table 1.11 Overview of the city of Brussels

Overview	
Continent	Europe
Country	Belgium
National GDP per capita in USD	47,561
National population	11,239,000
City population	1,180,000
Institution/City	Brussels
Department responsible for remote sensing data and resulting products	Brussels Regional Informatics Centre (BRIC)
Administrative area in km^2	161
Number of used remote sensing platforms	3
WebGis Service	Yes
Services like Google Street View, Google Maps or Bing Maps	Yes
Considering the importance of RS within public authorities as	Increasing

- Aerial imagery: used for orthophoto maps and aerial photogrammetry (2D, 3D cadaster)
- LiDAR
- Terrestrial data.

Aerial imagery is used to produce orthophoto maps and for aerial photogrammetry. Photogrammetry specifically concerns the updating of buildings in 2D and 3D as well as the updating of trees in roads. The terrestrial topography is mainly used for road surveys. The dates of acquisition in the past and the GSD for each acquired RS product are as follows:

- 1996 (GSD 10 cm)
- 1999 (GSD 10 cm)
- 2004 (GSD 10 cm)
- 2009 (GSD 10 cm)
- 2012 (GSD 7,5 cm)
- 2014 (GSD 7,5 cm)
- 2015 (GSD 7,5 cm)
- 2016 (GSD 7,5 cm)
- 2017 (GSD 5 cm)
- 2018 (GSD 5 cm)
- 2019 (GSD 5 cm) flight scheduled for April 2019.

Generally, annual acquisition flights are foreseen. However, the number of acquisitions and the respective planning also relates to the department related budgets. A new open call for tender will be launched for flights after 2019. This tender will last 4 years. Terrestrial surveying markets have been organized since 2007. Additionally, a new public tendering is being prepared for terrestrial topography, mobile mapping and indoor mapping. This tender will last for 4 years and will be accessible to the partners of the BRIC. A particularity related to mobile and indoor mapping is that in the first phase of these activities, there is a six-month proof of concept to validate the technologies. Aerial photography, photogrammetry and terrestrial topographic surveys have been outsourced to private companies in the context of public procurement.

As such, the BRIC does not have the monopoly for carrying out cartographic operations. Each organization in the Brussels Region can carry out its own work and, thus, acquire RS data on their own. For example, the organization that deals with environmental management in the Brussels Region has already acquired infrared orthophotos for a specific project. The public markets organized by the BRIC are accessible to organizations in the Brussels region. This is the case for the terrestrial surveying market. A municipality can call on the BRIC subcontractor to carry out terrestrial topographic surveys, avoiding the municipality making its own public markets. Since 2014, the BRIC has been organizing annual digital color aerial photography. The shots are mainly used to make orthophoto maps and to photogrammetrically update 2D and 3D buildings.

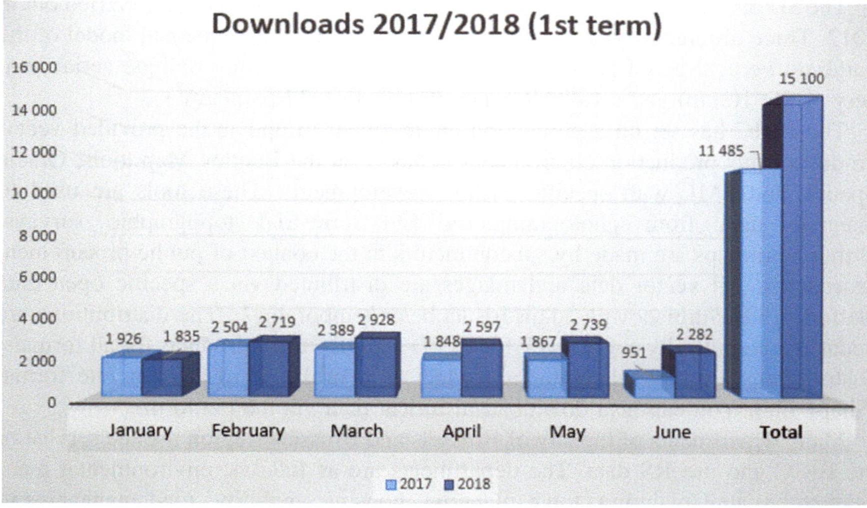

Fig. 1.2 Download statistics for the first half-year of the city of Brussels. *Source* CIRB

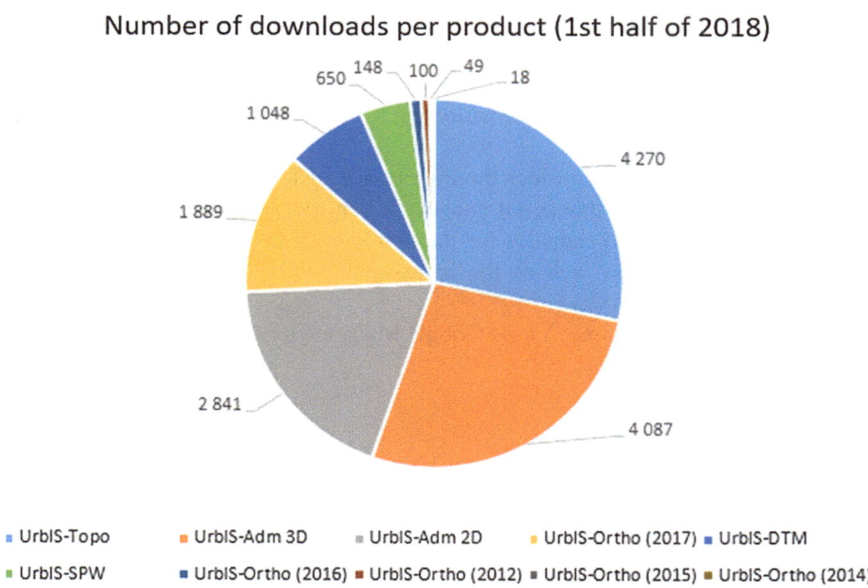

Fig. 1.3 Download statistics for the first half-year of the city of Brussels; shares of each product. *Source* CIRB

The 3D buildings were built on the basis of air operations that were carried out in 2012. Three different flights were used to build the three-dimensional model of the building: vertical aerial photography (GSD 7.5 cm), shooting oblique aerial imagery (GSD 10 cm) and a LiDAR frame (Resolution 32 points/m^2).

The BRIC has set up a production environment to update the provided vector products. This production environment is based on the Bentley Map tools, Oracle Spatial, and FME with specific Python developments. These tools are used to integrate data from photogrammetry and terrestrial topographic surveys. Orthophoto maps are made by subcontractors in the context of public procurement procedures. All vector data and images are distributed via a specific open data platform (http://urbisdownload.gis.irisnet.be/en/temporality/). The distributions are updated automatically every week for all 2D and 3D vector products for all formats. Distributions are done daily for a product (e.g. urbIS-Adm) and for one format (Shape file). You can also download historical data via the platform.

Many departments of the city of Brussels and Brussels Region use the service of the BRIC and use RS data. The departments are as follows: environmental management, spatial planning, town planning, housing, mobility, road management, waste collection, security, police and others. These departments and institutions generally use the catalogue of UrblS products. UrbIS®© comprises a set of geographical databases of the Brussels-Capital Region and online services (CIRB 2018).

One can certainly find indications on the Brussels Environment websites: https://environnement.brussels/doc and BruGIS https://mybrugis.irisnet.be/brugis/ about which departments use RS data and for which tasks; besides this information and user statistics (Fig. 1.2) about the downloaded RS data and resulting products, there is no specific documentation.

RS experts of Brussels consider the importance of RS within public authorities as increasing. They argue that aerial imagery is very popular, which is documented by the monthly download statistics for the first semester with a comparison to the same period last year, as well as the breakdown of downloads by product and format (Fig. 1.3).

Services like Google Street View, Google Maps are used, especially Street View are widely used.

1.4.1.3 France

In the city of Strasbourg (Table 1.12), the Service of Geomatics and Territorial Knowledge (French: Service Géomatique et connaissance du territoire) plays the main role in the acquisition of RS data and resulting products. The Service of Geomatics and Territorial Knowledge (SGTK) are part of the Management of Urbanism and Territory (French: Direction Urbanisme et Territoires). The SGTK has the lead for purchasing RS data and resulting products within the city administration of Strasbourg and, thus, functions as the interface for other municipal departments and divisions. RS data and resulting products used and provided by the SGTK are as follows:

Table 1.12 Overview of the city of Strasbourg

Overview	
Continent	Europe
Country	France
National GDP per capita in USD	42,779
National population	66,695,000
City population	280,000
Institution/City	Strasbourg
Department responsible for remote sensing data and resulting products	Service of Geomatics and Territorial Knowledge
Administrative Area in km^2	78
Number of used remote sensing platforms	4
WebGis Service	Yes
Services like Google Street View, Google Maps or Bing Maps	Yes
Considering the importance of RS within public authorities as	Increasing

- Aerial imagery,
- LiDAR,
- Satellite imagery: mainly high-resolution satellite imagery,
- Terrestrial laser scan data.

Data management and the distribution of RS data and resulting products are organized by means of an intranet platform, web forms, and can also be acquired by personal requests. Resulting products of the used RS data are as follows: orthophoto, 3D textured photo mesh, point clouds, raster grids of elevations, DTM, and vegetation layers.

The following departments of the city of Strasbourg use the services of the SGTK (Table 1.13).

The departments and divisions presented in Table 1.13 generally use the orthophoto, 3D data and satellite imagery for their tasks. Tasks RS data are used for are as follows (Table 1.14).

Besides this, there is no specific documentation regarding which city departments/divisions use RS data nor for which tasks or activities they are used. Requests for RS data and resulting products mainly come from communal areas; there are only a few requests from private entities.

There are web portals from which RS data and resulting products can be acquired:

1. http://www.cartotheque.strasbourg.eu/,
2. http://www.sig.strasbourg.eu/index.php?page=opendata
3. http://3d.strasbourg.eu/.

Table 1.13 Top departments and divisions that use RS data and resulting products

Departments and divisions
Spatial Planning
Building Authority
Heritage Building Unit
Division of Economic Development and Attractivity
Service for Prevention of Environmental Issues
Service for Green Spaces and Nature

Table 1.14 Tasks and activities RS data are used for

Tasks and activities
Production and maintenance of the tree cadaster
Production and maintenance of the photovoltaic cadaster
Environmental studies
Project simulations and landscape insertions
Mosquito eradication
Urban renovation and planning
Building insertion for architecture contests

RS data are provided free of charge in the form of downloadable data. However, if special products are requested that demand further processing, delivery costs may be charged. RS experts of Strasbourg consider the importance of RS within public authorities as increasing. They argue that the demand and the use of spatial datasets are generally growing both intercommunal and external, respectively.

Services like Google Street View, Google Maps or Bing Maps are hardly used.

1.4.1.4 Germany

For the city of Munich (Table 1.15), the Municipal Department for Geodata Services (German: Kommunalreferat GeodatenService) plays the main role in the acquisition of RS data; the Municipal Department for Geodata Services (MDGS) is the central point for other departments and divisions that wish to acquire RS data and resulting products. Other departments may then process these data internally.

High-resolution aerial imagery is acquired in 2-year time intervals—alternating between seasons with abundant vegetation and without vegetation. No other RS platforms or systems are used within the MDGS. Other departments have applied terrestrial laser scan systems, and the produced data are available. RS experts from the MDGS stated that there was interest in Sentinel-2A imagery because of the platform's temporal resolution. However, these data are not used within the city administration.

Data management and distribution are organized as follows: internally, data are provided through an intranet (Geoinfoweb), where staff may visualize geodata such

Table 1.15 Overview of the city of Munich

Overview	
Continent	Europe
Country	Germany
National GDP per capita in USD	50,715
National population	82,665,600
City population	1,410,000
Institution/City	Munich
Department responsible for remote sensing data and resulting products	Municipal Department for Geodata Services
Administrative area in km^2	310
Number of used remote sensing platforms	1
WebGis Service	Yes
Services like Google Street View, Google Maps or Bing Maps	No
Considering the importance of RS within public authorities as	Increasing

as aerial imagery. Externally, RS data in the form of, for example, thematic maps, may be visualized via a particular platform (maps.muenchen.de). The RS data and resulting products used within the city administration of Munich are as follows:

- Aerial imagery: orthophoto and true -orthophoto,
- DSM: available as raster dataset or point cloud derived from aerial imagery by means of stereo matching,
- DEM: derived by means of stereo matching of aerial imagery,
- Aerial imagery: up-to-date and historical imagery (back to 1925).

There is documentation regarding which city departments/divisions use RS data and which tasks or activities they are used for. The documentation was last conducted by means of a user evaluation in the year 2015. The following departments of the city of Munich use the services of the MDGS and, thus, RS data and resulting products, respectively (Table 1.16).

Table 1.16 Departments and divisions that use RS data and resulting products, sorted based on user importance

Departments and divisions
Building Authority
Dpt. of the Local Government
Dpt. of Urban Planning and Building Regulations
Directorate of the City of Munich
Dpt. of Health and Environment
Dpt. of Education and Sports
Dpt. of Public Order
Dpt. of Culture
Dpt. of Work and Economy

60–70% of internal customers are interested in aerial imagery and orthophotos because these data allow the user to gain a quick overview of a specific (spatial) situation. Plots may easily be measured and, thus, save costs by omitting on-site inspections. Illegal constructions may be identified without on-site inspections. In addition, aerial imagery is used for the evaluation of the tree vitality, in the course of preparation for traffic relevant interventions and for traffic information in general. The comparison of aerial imagery of two time-steps is used to monitor redensification or urban growth in general. The Department of Health and Environment, for example, is interested in the percentage of the total impervious surface or the assessment and distribution of green areas within the city. Biotopes are mapped and assessed by means of RS data.

Approximately 80% of all requests for RS data come from the administration (internal requests); about 20% of all request come from private entities. The spatial resolution of up-to-date aerial imagery for internal use is 10 cm and 20 cm for private users. Citizens are often interested in up-to-date aerial imagery or historic aerial imagery; companies request aerial imagery for the whole city or parts of the city DSM and DEM data is similarly requested. Externally, handling expenses are charged (e.g. for printing), whereas internal handling charges (buy & sell) do not exist anymore.

In January 2017, the platform opengov-muenchen.de was set up. Through this web platform, RS data, such as aerial imagery with 60 cm spatial resolution, are provided free of charge. Besides this, WMS and WMTS services are provided, which can also be used from outside Germany. RS experts of the MDGS consider the importance of RS and resulting products within public authorities as increasing. They justify this with the enhanced data quality that continuously increases, the rising amount of data, with the improved algorithms, and the increasing awareness within the city administration for this kind of data. In the context of the increasing awareness of RS data, there are plans to establish a core competence within the city administration of Munich merely dealing with the topic RS.

Services like Google Street View are not used within the MDGS. This fact is accounted for the general form of these data, the data quality, and the not documented quality assurance that actually impedes its usage.

In comparison to Munich, Erfurt (Table 1.17) is a smaller city and represents a typical provincial capital in Germany. Within the city of Erfurt, the provincial capital of Thüringen, the Department of Geoinformation and Land Division (German: Amt für Geoinformation und Bodenordnung) plays the main role in the acquisition of RS data. However, most of the RS data are acquired by the provincial government and its respective department, the Thuringian state office for survey and geoinformation (German: Landesamt für Vermessung und Geoinformation Thüringen). By means of its open government data platform (www.geoportal-th.de), the state of Thuringian provides RS data such as orthophotos or the digital elevation model. Within the city of Erfurt, city departments also derive their own products from RS data, e.g. the Gardening and Cemetery Department produces thematic maps. These produced

Table 1.17 Overview of the city of Erfurt

Overview	
Continent	Europe
Country	Germany
National GDP per capita in USD	50,715
National population	82,665,600
City population	210,000
Institution/City	Erfurt
Department responsible for remote sensing data and resulting products	Department of Geoinformation and Land Division
Administrative area in km^2	269
Number of used remote sensing platforms	2
WebGis Service	Yes
Services like Google Street View, Google Maps or Bing Maps	Yes
Considering the importance of RS within public authorities as	Increasing

maps are then transferred to the Department of Geoinformation and Land Division (DGLD) of the city of Erfurt, where they are stored in a city-wide geodatabase.

Within the city of Erfurt, aerial imagery is operationally used. LiDAR data have been used for individual projects, but not systematically. In the future, 3D laser scan data shall be used operationally. Neither satellite imagery nor radar data or terrestrial laser scan data are used. Other departments do not use satellite imagery justifying their decision that within densely built-up urban areas, the geometric resolution of optical satellite imagery is too low.

Internal data management and distribution are organized by means of an (intranet-) geodatabase that includes a geodata viewer. The city departments have different rights in access and use for data layers; i.e. the Department of Animal Health and Foodstuffs Control cannot access or view thematic maps of the Gardening and Cemetery Department. A public version of the geodata viewer exists that works similarly to the open data portals of other municipalities. Online forms for access or provision of RS data do not exist within the city of Erfurt but are used by the Free State of Thuringian.

RS data and resulting products used within the city administration of Erfurt are the orthophoto and thematic maps, such as the city base map (German: Stadtgrundkarte). The DGLD does not produce DEMs or DSMs; these RS products are produced and provided by the Free State of Thuringian. Until now, there is no 3D city model, but there are plans for its production, which will be done inhouse or contracted out to a private company. In the future, orthophotos may be acquired by the DGLD.

Almost all city departments use RS data and resulting products such as the city base map or specific thematic maps, e.g. the municipal office, the civil construction authority, the Department of Gardening and Cemetery (tree cadaster) or the Department of Culture.

There has been documentation regarding which city departments/divisions use RS data or for which tasks or activities. A user evaluation was conducted, where city departments were asked which objects or urban artefacts are missing and should be depicted in the city base map. For RS data and resulting products that are provided by the Free State of Thuringian (e.g. DEM, DSM etc.), there is no documentation.

As mentioned before, many requests for RS derived data sets, such as thematic maps, come from other departments. However, private companies also request thematic maps or layers. The intercommunal provision of RS data and resulting products is free of charge. For private entities, costs for planning, processing, and printing may occur. These (possible) costs are listed in a remuneration table that guarantees transparency to the purchaser. There is an internal (non-public) data portal: SVE GIS, as well as an external publicly accessible geodata portal of the city of Erfurt: http://geoportal.erfurt.de/gp/de/index.html.

RS experts of the DGLD consider the importance of RS and resulting products within public authorities as increasing. Services like Google Maps or Bing Maps have been used to some extent by the city departments when orthophotos showed limited usability. In the past, Google Street View has been partly used.

1.4.1.5 Hungary

In Hungary (Table 1.18), RS data and resulting products are used for urban planning, regional planning, development of smart city directives, development of templates and directives for modeling and architecture. However, the competences regarding acquisition and distribution of RS data are linked to different departments and institutes. The two main departments and institutes in Hungary that operationally use and provide RS data for planning purposes are the Department of Geodesy, Remote Sensing and Land Offices (former FÖMI: Institute of Geodesy Cartography and Remote Sensing), which is organized under the Government Office of the Capital City Budapest, and the Lechner Knowledge Center (Lechner Tudásközpont), which is a non-profit company owned by the prime ministry of Hungary. The Department of Geodesy, Remote Sensing and Land Offices (DGRSLO) is considered as the most important institute of Hungarian remote sensing.

I will start with the presentation of the Lechner Knowledge Center (LKC) that is responsible for supporting the Hungarian government in the topic of architecture and regional planning. The LKC is responsible for the regional and settlement planning activities of each municipality; they also develop regional plans for important areas, develop smart city directives, create templates and directives in BIM (Building Information Modeling) for modeling and architecture in general. In a nutshell, every Hungarian settlement regional plan is controlled by means of RS data used by the LKC. If the LKC has to acquire official RS data and resulting

Table 1.18 Overview of the Republic of Hungary

Overview	
Continent	Europe
Country	Hungary
National GDP per capita in USD	28,375
National population	9,690,000
City population	–
Institution/City	Republic of Hungary
Department responsible for remote sensing data and resulting products	Department of Geodesy, Remote Sensing and Land Offices, Lechner Knowledge Center
Administrative area in km^2	93,000
Number of used remote sensing platforms	4
WebGis Service	Yes
Services like Google Street View, Google Maps or Bing Maps	Yes
Considering the importance of RS within public authorities as	Increasing

products (i.e. something approved by law—e.g. the official terrain model) the DGRSLO has to be contacted; if official datasets are not mandatory, RS experts of the LKC acquire their own data by means of a UAV or use free datasets such as Landsat or Sentinel satellite imagery. In addition to UAV data and high-resolution satellite imagery, terrestrial laser scan data are used. The RS experts of the LKC state RS data is crucial for the following tasks:

- Regional and national development,
- NDVI derived biological activity, which is used for monitoring settlement developments,
- Green infrastructure is mapped for smart city analysis.

Data management and distribution are organized by means of an intranet. Generally, these data are used for control, monitoring and integration in GI systems, for building survey and regional studies in particular. Within the LKC, there is no documentation regarding which city departments/divisions use RS data nor for which tasks or activities they are used; UAV data are requested only. Data are provided free of charge for official public tasks; for all other requests, data are provided at the usual market prices. There is no web portal of the LKC from which RS data or resulting products can be acquired. RS experts of the LKC consider the need for and the use of spatial information rapidly as increasing. Services like Google Street View or Google Maps are used by almost all city administrations.

In contrast to the LKC, the DGRLSO mostly functions as an interface and first point of contact for the provision of RS data and resulting products. Besides this, the DGRLSO is responsible for the Hungarian Land Parcel Identification System (LPIS), a component of the IACS (Integrated Administration and Control System) for the CAP (Common Agricultural Policy) of the European Union. Although many municipal departments of bigger cities (around or above 80 thousand inhabitants) do use remote sensing data for various purposes, the DGRLSO is not an exclusive point of contact. RS data and resulting products used and by the DGRLSO are as follows:

- Aerial imagery,
- Orthophotos,
- LiDAR data,
- Terrestrial laser scan data,
- Medium-high resolution (HR) satellite imagery: TM7-8, SPOT6-7, RapidEye, Sentinel1-2, DMC, MODIS, THEOS, IRS-LISS, AWIFS, and SPOT4-5 (for internal use only),
- Very high-resolution (VHR) satellite imagery: WorldView1-4, GeoEye, Quickbird, Ikonos, and Pleiades (for internal use only).

For the Hungarian LPIS, the latest orthophotos are used. For the Control with Remote Sensing program (CwRS), the satellite data are acquired from the Joint Research Center (JRC), which is the European Commission's science and knowledge service. This satellite imagery is acquired for the CwRS for internal use only and is not provided to third parties as such. Aerial imagery, provided by the DGRLSO, is mainly used in cities with over 10–15 thousand inhabitants. Orthophotos are more frequently used than VHR satellites. LiDAR or terrestrial scan data are nowadays spreading among municipalities of 100,000, but the use of such systems and platforms has just started. Until now, there is no relevant know-how regarding how to comprehensively use these data; so far there is no regular use and broader experience in use within municipalities. RS experts of the DGRLSO consider the rising amount of these data commercially motivated as, until now and generally within municipalities, there is no real knowledge and capacity to use and analyze such data.

Data is managed and distributed by means of a geo shop website of the DGRLSO. The procurement of aerial photos works through personal request. Larger towns have intranet platforms; however, only a few are publicly available. The orthophoto, DEM, National Land Cover Database (CLC50) and DSM are the only products derived from RS data. Municipalities use these datasets for town management and monitoring of building regulations. The capital city (Budapest) and the biggest town (Debrecen) use a Green Space Intensity (GSI) product derived from NDVI of satellites (Landsat etc.) for green space monitoring and use LST (Land Surface Temperature) average values to analyse heat island effects in the settlement structure of urban areas.

There is no documentation regarding which city departments/divisions use RS data nor for which tasks or activities they are used. Mostly, requests for RS data come from communal areas. Besides this, private entities rarely request RS data. Older orthophotos (from the year 2000–2008) are free of charge. Other RS data are provided at a charge, even to other public entities. Mainly the public authorities want to buy RS data covering larger areas. The DGRLSO hosts a data providing office where local governments can buy raster or vector data.

RS experts of the DGRLSO consider the importance of RS within public authorities as increasing. Services like Google Street View, Google Maps or Bing Maps are used within city departments. However, they are not used as components for official tools, maps, certificates or any other (data) products provided by the DGRLSO.

1.4.1.6 Iceland

In the city of Reykjavik (Table 1.19), the Geoinformation Department (GD) plays the main role in the acquisition of aerial imagery in particular and RS data in general. Within the Reykjavik Municipality, the GD uses aerial imagery and terrestrial data. Other departments or divisions are able to take action and acquire RS data on their own. However, the GD is usually the first point of contact. Within the Municipality of Reykjavik, base maps, 3D city models, DEN, DSM and other GIS datasets are products of the GD. RS data and resulting products used and provided by the GD are as follows:

- Aerial imagery: acquired from low altitude flights (below or equal to 1000 m), 1–2 years' time interval.

Data management and distribution functions by means of an intranet and internet (web services); additionally, RS data are provided on personal requests. RS data are provided free of charge. There is a web portal (https://borgarvefsja.reykjavik.is/borgarvefsja/) where, for example, aerial imagery can be visualized; further, there is free access to all vector data through the web portal. From the aerial imagery,

Table 1.19 Overview of the city of Reykjavik

Overview	
Continent	Europe
Country	Iceland
National GDP per capita in USD	53,518
National population	350,000
City population	120,000
Institution/City	Reykjavik
Department responsible for remote sensing data and resulting products	Geoinformation Department
Administrative area in km^2	273
Number of used remote sensing platforms	1
WebGis Service	Yes
Services like Google Street View, Google Maps or Bing Maps	Yes
Considering the importance of RS within public authorities as	Remaining steady

orthophotos and base maps are produced within the GD. All departments and divisions use the service of the GD to a different degree. The main users are as follows: The Environment and Planning Department, the Office of Property Management and Economic Development, Reykjavik Energy, the Construction Department, the Public Works Department, the Education and Leisure Department, the Welfare Department and the City Parking Department.

There is no documentation regarding which city departments/divisions use RS data nor for which tasks or activities they are used. Nevertheless, the RS experts of Reykjavik state that RS data and resulting products are used, amongst others, for tasks in spatial planning, project works, construction works, demographic analysis, public works, presentations, meetings, and for monitoring of all kind of information.

RS experts of the GD consider that the importance of RS within public authorities has remained steady throughout the last years. Services like Google Street View, Google Maps or Bing Maps are used by the DG and other city departments.

1.4.1.7 Italy

In the city of Naples (Table 1.20), the Administration for Planning and Territory Management (Italian: La Direzione centrale pianificazione e gestione del territorio) plays the main role in the acquisition of RS data. The RS experts of Naples, of the Central Administration for Planning and Territory Management (DCPT), refer to the respective document on duties and functions of the municipal organization (Italian: Direttore Generale n.38/2012), in point 21 "affari generali". In this document, it is declared that the DCPT handles the task of acquisition of cartographic

Table 1.20 Overview of the city of Naples

Overview	
Continent	Europe
Country	Italy
National GDP per capita in USD	39,817
National population	59,433,744
City population	970,000
Institution/City	Naples
Department responsible for remote sensing data and resulting products	Administration for Planning and Territory Management
Administrative area in km^2	117
Number of used remote sensing platforms	2
WebGis Service	Yes
Services like Google Street View, Google Maps or Bing Maps	No
Considering the importance of RS within public authorities as	Increasing

material and the internal distribution. However, the experts indicate that the DCPT is not the exclusive point of contact since all departments are free to acquire data with independent means. RS data and resulting products used and provided by the DCPT are as follows:

- Aerial imagery,
- LiDAR data: acquired by the metropolitan city of Naples, which is an administrative division that forms the middle level of the territorial authorities of Italy together with the provinces.

Satellite imagery has not been used in recent years. Nevertheless, as part of European projects (e.g. Clarity) or as part of agreements between administrations (e.g. with the National Institute of Geophysics and Volcanology 'Istituto nazionale di Geofisica e Vulcanaologia'- INGV: Osservatorio Vesuviano), satellite data have been used. Research activity on requisites for climate services (Clarity project) will comprise use of satellite imagery. Additionally, LiDAR data has been used to estimate volumetric data from the DSM and DTM.

Data management and distribution functions by means of personal requests. Mainly the Department of Environment and Protection of Land and Sea (Italian: Direzione Ambiente, Tutela del territorio e del mare of Comune di Napoli) uses the service of the DCPT for tasks in the climate research project mentioned beforehand. In the future, it is planned to use satellite imagery for cadastre activities. Besides this. there is no documentation regarding which city departments/divisions use RS data nor for which tasks or activities they are used. RS data have not been acquired for specific purposes as regional authorities distribute datasets, generally free of charge, at a scale of 1:5000, which support planning and infrastructure activities.

The DCPT has free access to regional satellite data and orthophoto imagery, but it is not entitled to distribute these RS data or resulting products to other external organizations. There is no web portal from which RS data or resulting products can be acquired.

RS experts of the DCPT consider that the importance of RS within public authorities is increasing but that balance constraints do not allow provision of extended datasets. They explicate that remote sensing is a data source of high value; it is usually processed by universities or other educational and scientific institutions that are linked to the municipal administration by contracts or common interest agreements. As such, there is no general agreement with providers like Google maps or Bing maps for the provision of additional RS data or respective services.

1.4.1.8 Norway

In Oslo (Table 1.21), the Mapping Department of the Agency for planning and building services of the city of Oslo (Norwegian: Plan- og bygningsetaten, Oslo Kommune) plays the main role in the acquisition of RS data. Other city departments use their own staff for tasks in urban planning and energy planning by means of RS

Table 1.21 Overview of the city of Oslo

Overview	
Continent	Europe
Country	Norway
National GDP per capita in USD	60,978
National population	5,006,470
City population	667,000
Institution/City	Oslo
Department responsible for remote sensing data and resulting products	Mapping Department of the Agency for planning and building services of the city of Oslo
Administrative area in km^2	454
Number of used remote sensing platforms	6
WebGis Service	Yes
Services like Google Street View, Google Maps or Bing Maps	Yes
Considering the importance of RS within public authorities as	Increasing

data; in this context, they also collaborate with private companies. However, the mapping department usually is their (other city departments) primary contact point, at least for assistance. Thus, if other departments/divisions of the City of Oslo acquire/purchase RS data, the Mapping department is the interface and the first and exclusive point of contact. Therefore, the Mapping Department assists other departments/divisions regarding their need for RS data. The demand for assistance generally varies from one year to another. In a few special cases, other departments/divisions of the City of Oslo acquire/purchase the data and/or information themselves. RS data and platforms used by the city administration of Oslo are as follows:

- Aerial imagery: vertical images for mapping purposes and for orthophoto production every year,
- Oblique images (2 years' time interval) for viewing and measuring building fronts and facades, normally covering the built-up area of the municipality,
- LiDAR: 3D laser scan (3 years' time interval) normally covering the built-up area of the municipality,
- Radar data has been used to monitor subsidence (InSAR),
- Terrestrial laser scan data (instruments mounted on a vehicle) test-wisely used, but not operationally,
- Use of street view imagery (Inovitas and Cyclomedia systems) test-wisely used for central parts of the built-up area, but not operationally,
- Satellite imagery will be considered to be used in the future (Copernicus program),
- Use of hyperspectral aerial imagery is planned.

The orthophoto, DEM, and DSM for the City of Oslo are managed by the Mapping Department. The Mapping Department also fulfils data management tasks. These tasks include the production of specific layers and maps, organization of the internal use through intranet and internet map solutions, and the distribution according to long-term agreements and delivery based on personal request. The RS data and some of the resulting products are, as part of the Inspire directive, made available through the national spatial data infrastructure (http://www.geonorge.no). In addition, aerial imagery and orthophoto projects are available through: http://www.norgeibilder.no, national and local DEM/DSM projects are available through http://www.hoydedata.no.

The products provided by the Mapping Department encompass: Aerial images and laser data, which are mainly used for photogrammetric mapping, 3D model-ling (both physical and digital models), and thematic mapping (risk management, urban flooding, water management and use of green area/potential for green roofs, solar energy, wind and noise analysis, environment/climate information analysis and products). The following departments of the city of Oslo use the services of the Mapping Department and, thus, RS data and resulting products, respectively (Table 1.22).

Documentation about which departments use RS data and for which tasks exists and is available upon personal request. Requests for RS data come from public entities as well as private entities (e.g. companies, extra-faculty research institutes, individuals, etc.). Access to RS data through web-viewers is free of charge. Additional costs may occur for printing maps, special services or for access and use of RS data based on downloaded information. RS experts of the city of Oslo consider the need for and the use of spatial information increasing. Services like Google Street View, Google Maps or Bing Maps are used to some extent by the city departments.

Table 1.22 Top departments and divisions that use RS data and resulting products

Departments and divisions
Emergency Planning Agency
Municipal Undertaking for Social Housing
Agency for Fire and Rescue Services
Cultural Heritage Management Office
Agency for Urban Environment
Agency for Real Estate and Urban Renewal
Cemeteries and Burials Agency
Tax Collection Office
Municipal Undertaking for Cultural and Sports Facilities
Municipal Undertaking for Social Service Buildings
Oslo Port Authority
Agency for Planning and Building Services
Agency for Waste Management
Municipal Undertaking for Educational Buildings and Property
Education Agency
Agency for Water and Wastewater Services

1.4.1.9 Poland

In the city of Warsaw (Table 1.23), the Geodesy and Cadaster Department (Polish: Biuro Geodezji i Katastru) plays the main role in the acquisition of RS data and resulting products on the local level. The Head Office of Geodesy and Cartography (Polish: Główny Urząd Geodezji i Kartografii) is another governmental institution that also acquires RS data and resulting products for Warsaw, but with lower parameters such as resolution or frequency. As such, in most cases, the Geodesy and Cadaster Department (GCD) is the only source of RS data, and functions as an interface for other municipal departments. The GCD is involved in preparing different public procurements regarding RS data acquisition. In 2015, for example, the Heritage Protection Department purchased aerial imagery of the Vistula River during a low level of the river for the sake of archaeological research. RS data and resulting products used and provided by the GCD are as follows:

- Aerial imagery,
- LiDAR,
- Hyperspectral data,
- Terrestrial laser scan data.

RS experts of the city of Warsaw consider the acquisition of high-resolution satellite imagery on demand in the near future, e.g. when needed for a specific project. Spatial data are shared among the departments of the Municipal Office of Warsaw (Polish: Urzędu Miasta Stołecznego Warszawa) through a database dedicated to spatial data only (vector and raster data). This database is called the Repository of Central Database of Spatial Data of the City of Warsaw (Polish: Centralna Baza Danych Przestrzennych Miasta Stołecznego Warszawy). There are also separate servers that enable access from other departments to the database.

Table 1.23 Overview of the city of Warsaw

Overview	
Continent	Europe
Country	Poland
National GDP per capita in USD	29,291
National population	37,115,000
City population	1,700,000
Institution/City	Warsaw
Department responsible for remote sensing data and resulting products	Geodesy and Cadaster Department
Administrative area in km^2	517
Number of used remote sensing platforms	4
WebGis Service	Yes
Services like Google Street View, Google Maps or Bing Maps	Yes (regularly)
Considering the importance of RS within public authorities as	Increasing

Besides orthophotos and aerial imagery in general, the GCD provides a 3D city model of the city of Warsaw and several public green space maps based on hyperspectral data and other RS data. These public green space maps were ordered by the central Municipal Office. The RS experts of Warsaw explicate that they are aiming at the acquisition of aerial imagery when there are foliage- and snow-free conditions (e.g. April). These RS data would allow for the production of a detailed city map that will contain a higher number of recognized spatial objects.

The following departments of the city of Warsaw use the services of the GCD (Table 1.24).

The municipal departments use RS data and resulting products to carry out their tasks in accordance with legal regulations and technical guidelines. Orthophotos are used, for example, for spatial planning purposes. Terrestrial laser scan data is used for traffic signs detection (management of road belt). Hyperspectral data are used for trees species and condition (e.g. vitality, stress) detection (management of green areas). Besides this, there is no specific documentation regarding which city departments/divisions use RS data nor for which tasks or activities they are used. For this study, several municipal departments were contacted by the GCD. The departments provided information regarding their use of RS data and resulting products:

- Center of Public Communication (http://ngo.um.warszawa.pl/): The Center of Public Communication uses public web services such as Google Maps to delineate project-related areas. Resulting maps are published on the website of GCD as a help for project promoters and to put them in contact with the residents of Warsaw. Google Street View images are also used to show changes in the urban space connected with the implementation of a certain project, e.g. how the area looked before and after the project implementation. Projects are published on the following website: https://twojbudzet.um.warszawa.pl/aktualnosci/wydarzenia/szczeg-owe-zasady-v-edycji-w-dzielnicach.
- Municipal Waste Management Department: this department uses Google Street View—e.g. evaluation of the accessibility for vehicles (e.g. trucks), verification of the names of objects from banners, addressing, verification of the location of gazebos and waste containers. Google Maps is used to acquire information from labels, e.g. marking the location of services/ companies.
- Public Safety & Crisis Management Department: this department uses various public services (e.g. Google Street View, Google Maps, Bing Maps, etc.) for the

Table 1.24 Top departments and divisions that use RS data and resulting products

Departments and divisions
Architecture & Spatial Planning Department
Environmental Protection Department
Infrastructure Department
Public Safety & Crisis Management Department
Road Investment & Maintenance Coordination Department
Heritage Protection Department

verification of building characteristics such as building height, number of stories and accessibility in general.

- Education Department: this department regularly uses orthophotos to estimate the capacity of residential school circuits (e.g. for the evaluation of the Local Development Plans for a given area). The city geoportal is frequently used. Public web services are also used, such as Google Earth, Google Maps, Google Street View, OpenStreetMap, Planet Explorer, Targeo, the Polish Geostatistical portal (Geoportal.gov.pl) and Bing Maps. The RS data and resulting products are used for very different tasks, including historical analyses, verification of building and development status, the production of maps, the analysis of heights, distances and areas of various facilities.
- Heritage Protection Department: this department uses orthophotos and 3D models, and free satellite imagery (e.g. Landsat missions, Sentinel missions) to monitor land development and its variability over time, e.g. to determine the time and extent of the demolition of an area or tree-cuts. The Heritage Protection Department also uses LiDAR data (classified point clouds) for their analysis of buildings and the natural environment. They further use RS data for tasks related to the protection and care of monuments in consideration of the entire cultural and historical heritage of the city of Warsaw. RS data are seen to be necessary to contribute to the municipal records of monuments (monitoring, updating). Services like Google Maps, Bing etc. are used regularly by the department to identify all types of monuments in the urban area. Historical data (e.g. aerial photographs, satellite imagery, 360-degree views) are stored and published as part of these activities and services. RS experts of the city of Warsaw state that these activities are important, for example, in terms of object transformation—expansion, reconstruction/transformation, and demolition. The Heritage Protection Department carries out 3D analysis, analysis of the terrain compared to the development and the shape of greenery, analysis of the volume of objects and visibility analysis.

Requests for RS data and resulting products mainly come from public entities, research- and university institutes. Most of the RS data and resulting products are purchased under public procurement. Data free of charge are provided by the Head Office of Geodesy and Cartography and encompass orthophotos and laser scanning data.

There is a map portal (http://www.mapa.um.warszawa.pl/mapaApp1/mapa?service=fast_mapa) where data can be visualized, the Central Database of Spatial Data of the Capital City of Warsaw. The webpage connected with the map portal also includes addresses of network services such as WMS, WFS, REST/SOAP:

- WMS (Web Map Service) available on http://wms.um.warszawa.pl/serwis,
- WFS (Web Feature Service) available on http://wfs.um.warszawa.pl/serwis,
- API: allows for embedding maps in the HTML page,
- REST/ SOAP.

The map portal also includes current aerial imagery and historical imagery. RS experts of Warsaw consider the importance of RS within public authorities as increasing. They explicate that the importance of these data has grown significantly over the last few years. They argue that RS data has become more accessible and more commonly used in tasks regarding the natural environment and the management of urban space. As such, they state that each year Warsaw publishes new orthophotos for the entire Warsaw area since 2010. The aerial imagery acquired lately now has an enhanced GSD (currently 8 cm). According to the RS experts of the city of Warsaw, the Municipal Office will publish oblique aerial imagery on a dedicated webpage next year.

1.4.1.10 Portugal

In the city of Lisbon (Table 1.25), the Department of Information Systems (Portuguese: Departamento de Sistemas de Informação)—through the Division of Georeferenced Information Management, shares the skills of acquiring the information with the Land Survey Division. The (spatial) information are provided to the municipal services in a centralized way. Although the Department of Information Systems (DSI) is the preferred contact point, other departments or service institutions can promote the acquisition for specific purposes. Mostly aerial images are used within the DSI. Further, recent satellite imagery is used, and 3D laser scan data are acquired for specific urban planning purposes. Data are available in a georeferenced format in a shared folder on a server with read access to all users of the municipal services. Some data are available for consultation to the general public in

Table 1.25 Overview of the city of Lisbon

Overview	
Continent	Europe
Country	Portugal
National GDP per capita in USD	32,199
National population	10,561,614
City population	550,000
Institution/City	Lisbon
Department responsible for remote sensing data and resulting products	Department of Information Systems
Administrative area in km^2	100
Number of used remote sensing platforms	3
WebGis Service	Yes
Services like Google Street View, Google Maps or Bing Maps	Yes
Considering the importance of RS within public authorities as	Increasing

a web application (Lisboa Interativa: http://lxi.cm-lisboa.pt/lxi/), and, in some circumstances (academic studies in restricted areas of the city, protocols with public entities, etc.), data are provided by completing an official form which is sent on request. At the moment, only orthophotography and DTM are provided by the DSI. As a basic cartography dataset, all departments use the orthophoto component. Land modeling is used very broadly, being one of the main datasets usable for tasks in urban planning, civil protection and the environment.

There is no documentation regarding which city departments/divisions use RS data nor for which tasks or activities they are used. Requests for RS data and resulting products mostly come from public entities but also from academia (e.g. universities). RS data is provided free of charge along with restrictions regarding their distribution. There is no web portal where RS data or resulting products can be acquired from. RS experts of the DSI consider the importance of RS within public authorities as increasing, based on their lower complexity and costs when compared to classic cartography. Services like Google Street View, Google Maps or Bing Maps are used periodically in the DSI and in other departments mainly through the use of the respective APIs in mobile applications that contain georeferenced information (mainly ESRI, Google Maps and Open Street Maps).

1.4.1.11 Spain

The Geoportal (GP) of the spatial data infrastructure of the Community of Madrid (Spanish: Geoportal de la Infraestructura de Datos Espaciales de la Comunidad de Madrid) plays the main role in the acquisition of RS data for this area (Table 1.26).

Table 1.26 Overview of the Community of Madrid

Overview	
Continent	Europe
Country	Spain
National GDP per capita in USD	38,091
National population	46,600,000
City population	6,550,000
Institution/City	Community of Madrid
Department responsible for remote sensing data and resulting products	Geoportal of the spatial data infrastructure of the Community of Madrid
Administrative area in km^2	8021
Number of used remote sensing platforms	2
WebGis Service	Yes
Services like Google Street View, Google Maps or Bing Maps	Yes
Considering the importance of RS within public authorities as	Increasing

Usually, the municipal authorities use the service of the GP. However, the Municipality of Madrid and the state-owned enterprise (owned by the Community of Madrid) Canal de Isabel II, which manages the water supplies for the region city of Madrid, may acquire remotely sensed imagery themselves. The GP uses aerial imagery (oblique also) and high-resolution satellite imagery. RS datasets and layers are mostly provided through personal requests (e.g. email) but the Community of Madrid is working on the implementation of a download area on their website. Nevertheless, to the RS experts of the GP, it appears that the online platform IDEM (Comunidad de Madrid 2018) is increasingly used. Besides the mentioned RS platforms and data, the GP also acquires LiDAR data and RGB imagery. Usually, departments dealing with cartography, urbanism and the environment use the service of the GP for tasks in urbanism and environmental control and protection (e.g. dumpsites).

There is no documentation regarding which city departments/divisions use RS data nor for which tasks or activities they are used. Besides municipalities, the universities and engineering consultants are important users of the GP's services. Data are provided free of charge, respecting the property rights. Until now, the web portal merely allows for visualization of aerial imagery, oblique imagery, satellite imagery, and the visualization of several GIS layers. RS experts of the GP consider the need for and the use of spatial information increasing in the medium-term. Services like Google Street View, Google Maps or Bing Maps are used by all city departments and municipalities.

1.4.1.12 United Kingdom

In the United Kingdom, the Ordnance Survey plays the main role in the acquisition of RS data and resulting products, and for the provision of vector data and basemaps. The Ordnance Survey (OS) is a limited company incorporated in England and Wales and plays a fundamental role in providing vector data. Their shares are 100% owned by the Secretary of State for Business, Energy & Industrial Strategy (BEIS) (OS 2018). The OS provides address data, postcode data, raster maps, transport network and vector topographical maps. They incorporate remotely sensed data in the regular update of their vector products. As a government-owned company, the OS provides RS data and resulting products to the public and private sector. In the framework of the Public Sector Mapping Agreement (PSMA) from the 1st of April 2011, the OS gives public sector organizations in England and Wales access to OS mapping products, including free access to several datasets of the OS.

Additionally, many city councils in the United Kingdom have their particular GIS/IT departments that are able to purchase RS data and resulting products independently from those offered by the OS. There is no legally binding requirement for city councils that RS data have to be purchased from the OS.

In the city of Barnsley (Table 1.27), the Information Technology division (IT) of the Barnsley Metropolitan Borough Council plays the main role in the acquisition

Table 1.27 Overview of the city of Barnsley

Overview	
Continent	Europe
Country	United Kingdom
National GDP per capita in USD	43,877
National population	63,181,775
City population	91,000 (city)
Institution/City	Barnsley
Department responsible for remote sensing data and resulting products	Information Technology division of the Barnsley Metropolitan Borough Council
Administrative area in km^2	329 (Barnsley Metropolitan Borough Council)
Number of used remote sensing platforms	2
WebGis Service	Yes
Services like Google Street View, Google Maps or Bing Maps	Yes
Considering the importance of RS within public authorities as	Increasing

of RS data and resulting products. Usually, other divisions contact the IT office for the acquisition of software or data. However, other divisions are free to acquire RS data on their own. RS data and resulting products are used within the Barnsley city council as follows:

- Aerial imagery: RGB, 0.125 m spatial resolution, available for the years 2002, 2009 and 2013,
- LiDAR: used once (2015), acquired from the Environment Agency which is a non-departmental public body of the United Kingdom Government's Department for Environment.

Aerial imagery is regularly used by the planning division (for both Local Development Framework and Development Control) for visual interpretation and digitizing of, for example, land use or land cover changes.

Aerial imagery is stored on a network share in tiled MrSID and JPG format. On a MapInfo site, all tiles are associated with the respective MapInfo metadata (.TAB file, functionally world file). There are also several 'seamless tables' (functionally image catalog) covering Barnsley and South Yorkshire in all three aerial image sets (2002, 2009, and 2013). The 2013 dataset is also provided as internal WMS. The main users of aerial imagery are the divisions responsible for urban planning, gazetteer management, highways management, tree preservation and public-facing maps. These departments use RS data and resulting products for planning control, planning enforcement, and the local development framework. Besides this, there is no direct documentation regarding which city departments/divisions use RS data nor for which tasks or activities they are used.

Requests for RS data and resulting products generally come from communal areas (mostly from within councils). During the last year, less than ten requests for geodata were counted from private entities and universities.

Until the 1st of April 2018, free RS data were quite limited for local governments. From the 1st of April (2018), aerial imagery is available to town-councils as part of the Aerial Photography Great Britain Agreement (APGB). This agreement is similar to the OS PSMA. The following products are available: 12.5 cm (most of GB) and 25 cm color imagery, 50 cm CIR, 2 m DSM and 5 m DTM. The IT plans to make these datasets internally available. A platform where RS data and resulting products are publicly available is http://data.gov.uk.

RS experts of the IT consider the importance of RS within public authorities as increasing because they perceive an increasing public demand for 3D models. Services like Google Maps (satellite), Street View and Bing Maps (aerial and bird's eye) are used by the gazetteer management and the planning division for reference.

1.4.2 North America

1.4.2.1 Canada

In the city of Calgary (Table 1.28), the Corporate Analytics & Innovation (CAI) business unit plays the main role in the acquisition of RS data and resulting products. The CAI provides operational, technical and strategic support to the City of Calgary and to its municipal departments and divisions. The CAI further

Table 1.28 Overview of the city of Calgary

Overview	
Continent	North America
Country	Canada
National GDP per capita in USD	46,378
National population	35,151,000
City population	1,160,000
Institution/City	Calgary
Department responsible for remote sensing data and resulting products	Corporate Analytics & Innovation business unit
Administrative area in km^2	825
Number of used remote sensing platforms	2
WebGis Service	Yes
Services like Google Street View, Google Maps or Bing Maps	Yes
Considering the importance of RS within public authorities as	Increasing

provides extensive geospatial products and data for the City of Calgary. The CAI mainly acquires aerial imagery; terrestrial data are less frequently acquired. Generally, aerial imagery is used within the CAI business unit. However, in some municipal projects, terrestrial laser scan data are also used. Resulting products such as the Orthophoto or the DSM are made available corporately on an image service. RS data and resulting products are distributed by means of an image service delivery, ESRI ArcGIS, or AGOL applications. RS experts of the City of Calgary state that data are extensively used by the following business units and departments: the Water Resources Business Unit, the Water services business unit, the Roads business unit, the Parks business unit, and the Transportation Department in general.

Besides this knowledge about the main intercommunal clients, there is no specific documentation regarding which city departments/business units use RS data nor for which tasks or activities they are used.

Requests for RS data come from communal areas (within public authorities) and also from private entities, such as companies, extra-faculty (non-university) research institutes, university institutes, and individuals. RS experts of the City of Calgary explicate that some RS data and resulting products are provided free of cost; for other datasets, a nominal fee is charged. The experts consider the importance of RS within public authorities as increasing. Services like Google Street View, Google Maps, and Bing Maps are used corporately within the City by some departments.

In the city of Toronto (Table 1.29), the Geospatial Competency Centre (GCC) is responsible for collecting annual city standard aerial photography. Other city divisions can collect remotely sensed data without the GCC. However, usually, other divisions contact the GCC for assistance on technical specifications and assistance with procurement. The City of Toronto uses a wide range of remotely sensed data; RS data and platforms that are used are as follows:

Table 1.29 Overview of the city of Toronto

Overview	
Continent	North America
Country	Canada
National GDP per capita in USD	46,378
National population	35,151,000
City population	2,810,000
Institution/City	Toronto
Department responsible for remote sensing data and resulting products	Geospatial Competency Centre
Administrative area in km^2	630
Number of used remote sensing platforms	6
WebGis Service	Yes
Services like Google Street View, Google Maps or Bing Maps	Yes
Considering the importance of RS within public authorities as	Increasing

- Four-band 6 cm aerial photography every year,
- Aerial LiDAR: last acquisition in the year 2015, planned acquisition at a particular interval.

Other remotely sensed data are being collected within different divisions (terrestrial LiDAR, satellite images, traffic cameras, and drone imagery). RS data management and distribution are organized depending on the source. The following data are made available by the GCC:

- Stereo imagery: Stereo images are processed and provided. The images are stored on a file server organized by folders denoting the year of collection. Geospatial photo center points are available for each year of imagery.
- Ortho imagery: The organization of orthophotos consists of three methods: (1) internal local file server (tiled images organized by year—only two most recent years); (2) internal corporate file server (compressed and images organized by year—all years available); (3) Web map service (WMS) stored and maintained in ArcServer©.
- Aerial LiDAR: The organization of LiDAR consists of three methods: (1) internal local file server (tiled images organized by year); (2) internal corporate file server (tiled images organized by year); (3) Web map service (WMS) stored and maintained in ArcServer©—only hillshade data is made available.

The internal and external data distribution differs as follows: Internally, City employees request access to the tile-based data directly from the GCC, which is granted. Employees are provided access to the file servers and can access them at will. WMS are made available to all employees. WMS are used within different applications. All external requests are made through the GCC.

The GCC has several different products available, such as topographic vector features (collected using stereo compilation from stereo aerial photography), ortho imagery, LiDAR (bare earth DEM, full feature DSM, slope, aspect, hill-shades and contours), and permeability vectors (derived from ortho imagery). The following departments of the City of Toronto represent the top divisions that use the services of the GCC and, thus, RS data and resulting products, respectively (Table 1.30).

However, there is no explicit documentation by means of customer surveys or requests regarding which city departments/divisions use RS data nor for which tasks or activities they are used. The GCC maintains a "Map Desk", which provides data service access for external organizations or individuals. Usually, universities,

Table 1.30 Top departments and divisions that use RS data and resulting products

Departments and divisions
City Planning
Parks Forestry & Recreation
Toronto Water
Transportation Services
Public Health
Engineer Construction Services

utility companies and consulting firms take advantage of this service. Some RS data are free of charge, whereas others are associated with costs. Web map services for ortho imagery is available for free on the City's Open Data site. In general, all data pro vided via the City's Open Data platform are free of charge; additionally, if certain datasets are used for a City of Toronto project, they are free of charge; otherwise such data incur costs. Web map services are available from the City's Open Data site. RS experts of the City of Toronto consider the importance of RS within public authorities as increasing. In addition to the existing city geospatial digital infrastructure, services like Google Street View, Google Maps or Bing Maps are used within the municipal organization.

1.4.2.2 Mexico

In Mexico City (Table 1.31), the Environmental and Land Management Office of the City of Mexico (PAOT) (Spanish: La Procuraduría Ambiental y del Ordenamiento Territorial de la Ciudad de México) acquires information obtained by remote sensors, such as satellite imagery or mosaics of aerial photography. According to the RS experts of the city of Mexico, this acquisition of RS data is not regularly conducted and does not respond to an annual program as such. RS data supports many institutions for the investigation of, for example, complaints. As such, RS data and resulting products constitute a part of the spatial information collection of the PAOT.

Table 1.31 Overview of Mexico City

Overview	
Continent	North America
Country	Mexico
National GDP per capita in USD	18,149
National population	113,580,528
City population	8,920,000
Institution/City	Mexico City
Department responsible for remote sensing data and resulting products	Environmental and Land Management Office of the City of Mexico
Administrative area in km^2	1,485
Number of used remote sensing platforms	3
WebGis Service	Yes
Services like Google Street View, Google Maps or Bing Maps	Yes
Considering the importance of RS within public authorities as	Increasing

It remains unclear to the author whether other departments take action and may acquire RS data on their own. The RS experts of the PAOT state that the PAOT usually acquires RS data for the city of Mexico in respect to the financial resources of the department. RS data and platforms used by the PAOT are as follows:

- Aerial imagery: image acquisition flight for the whole urban area of Mexico City for the years: 1986, 1995 and 1996; 1980–2014 image acquisitions of subareas,
- Satellite imagery: QuickBird image mosaic of the year 2008, a mosaic of World View images of the year 2015,
- LiDAR data: point cloud data that covers the nature conservation area called "Ejidos de Xochimilco y San Gregorío Atlapulco", acquisition date: November 2014.

According to the RS experts of the PAOT, two sub-divisions of the PAOT are mainly responsible for the preparation of studies and reports on environmental protection and land use: the subdivision of Land (Spanish: Subprocuraduría del Ordenamiento Territorial) and the subdivision of Environmental Planning and the Protection of Animals (Spanish: Subprocuraduría Ambiental, de Protección y Bienestar a los Animales). The experts state that RS data and resulting products are concentrated in these areas and institutions.

The PAOT has a collection of information obtained from RS data, such as mosaics of satellite imagery and aerial imagery. Other RS data and resulting products are handled by the Ministry of Urban Development and Housing. Information on existing datasets, which are mostly related to land use information, are available on their respective website (http://ciudadmx.cdmx.oob.mx:8080/seduvi/).

The Environmental and Territorial Planning Office of Mexico City uses RS data and spatial data in general to prepare technical documents and to contribute to other administrative tasks. Some of these data are integrated into the GIS web platform of the PAOT (SIG-PAOT): http://200.38.34.15:8008/mapguide/sig/siginterno.php, which is available on the official website: http://www.paot.org.mx/. RS data and resulting products provided through the websites mentioned beforehand are free of charge. Besides this, there is no explicit documentation by means of customer surveys or requests regarding which city departments/divisions use RS data nor for which tasks or activities they are used. RS experts of the PAOT explicate that such information is dispersed and may only be accessed by individual consultations of the respective departments and divisions.

Requests for RS data and resulting products mostly come from communal areas (within public authorities) but also from academia (e.g. universities).

RS experts of the PAOT consider the importance of RS within public authorities as increasing. They underpin this with the fact that the information generated from RS data is of great importance for the analysis of the territory of Mexico City. Services like Google Street View and Google Maps are used within the PAOT day-to-day, and the extracted information are integrated into documents and reports.

1.4.2.3 United States of America

In the city of Denver (Table 1.32), geospatialDENVER takes care of any enterprise type of project or data acquisition. GeospatialDENVER is the umbrella term or

Table 1.32 Overview of the city of Denver

Overview	
Continent	North America
Country	United States of America
National GDP per capita in USD	59,532
National population	308,745,538
City population	680,000
Institution/City	Denver
Department responsible for remote sensing data and resulting products	geospatialDENVER (on behalf of the Technology Services—GIS Department)
Administrative area in km^2	401
Number of used remote sensing platforms	3
WebGis Service	Yes
Services like Google Street View, Google Maps or Bing Maps	Yes
Considering the importance of RS within public authorities as	Increasing

organization, respectively, for the Technology Services—GIS Department (TS-GIS) and its services. It describes their whole program (Agencies using GIS + Technology Services GIS or the Enterprise GIS Group in Technology Services). TS-GIS has a standardized aerial photography program that encompasses requirements from all city agencies. Every two years, the TS-GIS department collaborates with the Denver Regional Council of Governments (DRCOG) and collects ten counties worth of aerial photography at various resolutions. This acquisition activity is called the Denver Regional Aerial Photography Project (DRAPP). Although TS-GIS plays the main role in the acquisition of RS data, sometimes city agencies buy historical data products, and some city departments have their own UAV fleet. RS data and resulting products used and provided by the TS-GIS are as follows:

- Aerial imagery: 4 bands, 2 years' time interval, various spatial resolutions,
- Oblique aerial photography: 2 years' time interval,
- Aerial LiDAR: 8 years' time interval, every two years: collection of planimetric features (building rooflines, edge of pavement, sidewalk, parking lots, driveways, etc.) as part of the DRAPP,
- Very high-resolution satellite imagery: an imagery streaming service of a well-known provider is installed for 30–60 cm pan and natural color satellite imagery.

Data management and the distribution of RS data and resulting products are organized as follows: The original .tiff or .jp2 files are stored on a shared drive. An image mosaic is created in-house. Then, the data are served out to the GIS users, and integrated applications are managed by means of an image server web service of a well-known provider. Special file formats such as .sid (compressed image file format) images are provided on and for external requests. The provision of RS data such as aerial imagery depends on its acquisition date: Only data before the last flight can be provided, i.e. RS data dating back to 2014 and data gathered in earlier flights are publicly available. In addition to the RS datasets mentioned above, there is also a dataset of 1-foot (equates: 30.48 cm) contour data that was generated from the LiDAR data. It is very often used for planning purposes. The planimetric datasets mentioned above are derived from the orthophotography.

Just about all 34 city agencies (e.g. Human Services, Public Safety, Community Planning and Development, Environmental Health, Parks and Recreation and others) of Denver use RS data that is provided by the TS-GIS department. There is an internal GIS web-based application called Map It Denver, which allows users to turn on aerial photography from different years (2016, 2014, 2012, 2010, 2008, 2006, 2004, 2002, 2000, 1995, 1956, 1933) as well as many other GIS layers.

There is documentation outlining which departments use RS data and the service of the TS-GIS department in general and for which tasks. To the author of this study, a supplement was presented that provides comprehensive information about which departments use RS data, for which purposes, and which RS data are explicitly beneficial for their work.

An excerpt of this document, which presents the tasks and activities RS data and resulting products are used for, is given as follows (Table 1.33).

Requests for RS data comes from communal areas (within public authorities) and also from private entities, such as companies, extra-faculty (non-university) research institutes, university institutes, etc. In general, aerial photography and

Table 1.33 Tasks and activities RS data are used for

Tasks and activities
Forestry
Planning
Policing
Addressing
Impervious surface mapping for e.g. wastewater management
Assessments
GIS feature collection for e.g. verification of topographic features in AutoCAD drawings
Environmental reviews
Identification of park assets
Monitoring of the redevelopment of brownfields properties
Assessment of storm water and ground water behaviour
Determination of drainage basin sizes and drainage basin flow directions
Virtual site visits
Aerial imagery serves as a form of ground-truthing of new spatial datasets

satellite imagery as paid for web services are seen to gain importance. There are a lot of requests for aerial photography. RS data are provided for free. Some aerial photography is publicly available on https://www.denvergov.org/opendata.

RS experts of the city of Denver consider the need for and the use of spatial information as increasing. The RS experts state that policy makers, especially, are in favor of aerial photography and have come to expect high quality, accurate, and current data. According to the RS experts of Denver, the city is experiencing an enormous growth boom, and the landscape is constantly changing. Therefore, historical aerial photography is becoming more important as planners want to preserve the character of Denver's neighborhoods. Further, they explicate that in the context of the recent building boom within Denver, a contract has been established with a provider of streaming aerial photo services. This company acquires aerial imagery at least two times a year.

Services like Google Street View, Google Maps or Bing Maps are used by many departments. The TS-GIS department uses Google Street View and Google Maps to perform research on tasks such as determining actual street names, address verification, etc.

For the County of Los Angeles (Table 1.34), the Los Angeles Chief Information Office (LACIO) is the lead agency for the Los Angeles Region Imagery Acquisition Consortium (LARIAC) Program. This program is a collaborative acquisition program for digital aerial imagery data for government entities that are part of Los Angeles County. The County is the lead agency, and other government agencies can join based on their interest and need.

Table 1.34 Overview of Los Angeles County

Overview	
Continent	North America
Country	United States of America
National GDP per capita in USD	59,532
National population	308,745,538
City population	9,820,000 (County)
Institution/City	Chief Information Office of Los Angeles County
Department responsible for remote sensing data and resulting products	Los Angeles Chief Information Office (lead agency for the Los Angeles Region Imagery Acquisition Consortium [LARIAC] Program)
Administrative area in km^2	12,310
Number of used remote sensing platforms	2
Web-Gis Service	Yes
Services like Google Street View, Google Maps or Bing Maps	Yes
Considering the importance of RS within public authorities as	Remaining steady

LARIAC participating agencies currently include 22 County departments, 30 municipalities, and four other public agencies. The RS experts of Los Angeles County consider this consortium as a model for regional cooperation that improves the quality, access, and cost-effectiveness of high-resolution digital aerial imagery (Franchino 2018). The staff members of the LACIO work with over 50 agencies to determine the remote sensing products, as well as the large array of derived products, that are part of the portfolio. County departments can acquire additional imagery on their own with the approval of the LACIO. Additionally, there is a separate quality assurance process to ensure data quality.

It is worth mentioning that the City of Los Angeles is a completely different political entity. Its departments (e.g. Department of Recreation and Parks) collaboratively fund the city's participation in LARIAC and need to go through their particular procurement rules when they wish to acquire additional RS data. The City of Los Angeles has a central GIS coordinator who, analogous to the person of the Los Angeles county organization LACIO, reviews all remote sensing data. Most procurements go through the Chief Executive Office of the city of Los Angeles for approvals. However, the city of Los Angeles and the respective departments are familiar with the LARIAC program and the provided datasets and have been a member since inception.

The County of Los Angeles uses remotely sensed data; the RS data and platforms used are as follows:

- Aerial imagery: Orthoimagery (including NIR) and oblique imagery at two resolutions: nominal 10 cm Ground Sample Distance (GSD) and nominal 28 cm GSD,
- Aerial LiDAR: last acquisitions in the years 2006 and 2015, planned acquisition on a regular $\sim$10-year interval; county DTM with up to 2 m spacing.

More information about the platforms used and about the specifications of the resulting RS data are provided on the LARIAC program website, e.g. http://egis3.lacounty.gov/dataportal/wp-content/uploads/2013/07/LARIAC4-Product-Guide-1.pdf.

Key data derived from the source imagery and LiDAR include DEM, DSM, DTM, building rooflines, a land cover dataset, an NDVI layer, and a solar radiation model.

Data management and the distribution of RS data is organized by means of hard disks that are provided, as well as web mapping services. Other requests are handled via email. Further, several online mapping systems are provided for the oblique imagery enabling access and include web development widgets.

All County departments (e.g. Fire Department, Department of Beaches & Harbours, Department of Parks and Recreation, Department of Regional Planning) and over 40 city, state and federal agencies, water companies, research universities, and local jurisdictions use the RS data and resulting products provided by the LACIO. The LACIO uses RS data for disaster planning, code enforcement, provision of basemaps, GIS data development, public works, public safety (law enforcement), sustainability, energy efficiency, and many other purposes.

There is no specific documentation regarding which city departments/divisions use RS data nor for which tasks or activities they are used. Primarily, requests for RS data and resulting products come from government entities. However, there are also requests from private entities, such as companies, extra-faculty (non-university) research institutes, university institutes, and individuals.

Lower quality (down sampled) RS data are provided free of charge through the partnership with the US Geological Survey. There is a web portal (https://egis3.laccunty.gov/dataportal/lariac/) which describes the products, yet, not all products can be acquired directly from there; contact information must be provided to obtain the data.

RS experts of the county of Los Angeles consider the need for and the use of spatial information as remaining steady. Services like Google Street View and Google are used for web development.

1.4.3 South America

1.4.3.1 Bolivia

In the city of La Paz (Table 1.35), the institutional network GeoBolivia plays the main role in the acquisition of RS data and resulting products. However, there is no binding legal framework that obligates other institutions to use GeoBolivia as their first point of contact or interface for the acquisition of RS data and resulting

Table 1.35 Overview of the city of La Paz

Overview	
Continent	South America
Country	Bolivia
National GDP per capita in USD	7,560
National population	11,307,000
City population	760,000
Institution/City	La Paz
Department responsible for remote sensing data and resulting products	GeoBolivia
Administrative area in km^2	2,012 (Municipio de la Paz)
Number of used remote sensing platforms	3
WebGis Service	Yes
Services like Google Street View, Google Maps or Bing Maps	Yes
Considering the importance of RS within public authorities as	Increasing

products. GeoBolivia, as such, is a computer platform that allows storing, searching and publishing geographic information of Bolivia through a web portal; from this web portal, "it is possible to have relevant, harmonized, and quality geographic information to support decision-making and the social, economic and environmental development of Bolivia" (Geobolivia 2018). RS data and resulting products used and provided by GeoBolivia are as follows:

- Aerial imagery,
- High-resolution satellite imagery,
- UAV: laser scan data.

Data management and the distribution of RS data is organized based on personal requests due to challenges in providing a broadband internet connection. RS experts of the city of La Paz state that many departments and municipalities use the service of the GeoBolivia. However, there is no documentation regarding which city departments/divisions use RS data nor for which specific tasks or activities they are used. The RS experts indicate that RS data and resulting products are used for disaster management, cadastre related operations, and for the identification of actual land use. GIS datasets and layers are provided free of charge via the web portal of GeoBolivia (http://geo.gob.bo/portal/) and the respective data catalog. RS experts of La Paz consider the need for and the use of spatial information as increasing; they base this opinion on the fact that all official reports must have a territorial base. Services like OpenStreetMap, Mapillary, and MapBox are used within the municipal organization.

1.4.3.2 Brazil

The Paulista Company of Metropolitan Planning SA (Portuguese: A Empresa Paulista de Planejamento Metropolitano SA) is responsible for the regional and metropolitan planning of the State of São Paulo (Table 1.36). The Paulista Company of Metropolitan Planning SA (Emplasa) is a public institution linked to the State Secretariat of the Civil House. Created in 1974 in the context of planning greater São Paulo, it now provides technical support to regional institutionalized units: Metropolitan Regions of São Paulo, Campinas, Baixada Santista, Vale do Paraíba and Litoral Norte and Sorocaba, and the Urban Agglomerations of Jundiaí and Piracicaba. These regions are part of the macro-region called Macrometropole Paulista (Portuguese: Macrometrópole Paulista); as such, the prefecture of São Paulo is a constitutional part of this macro-region and the federal state of São Paulo. To complete the picture, the federal state of São Paulo is one of the 26 states of the Federative Republic of Brazil.

In the federal state of São Paulo, Emplasa elaborates and subsidizes the Government of the State of São Paulo in the implementation of public policies and integrated urban and regional development projects. It also carries out diverse

Table 1.36 Overview of the state of São Paulo

Overview	
Continent	South America
Country	Brazil
National GDP per capita in USD	15,484
National population	190,700,000
Federal population	45,600,000
Institution/City	State of São Paulo
Department responsible for remote sensing data and resulting products	The Paulista Company of Metropolitan Planning SA (Emplasa)
Administrative area in km^2	248,222
Number of used remote sensing platforms	2
WebGis Service	Yes
Services like Google Street View, Google Maps or Bing Maps	Yes
Considering the importance of RS within public authorities as	–

studies and provides cartographic products, geospatial information systems and technical knowledge about metropolitan planning to the public.

Since 2015, Emplasa has been responsible for the collection and the state cartographic production (SCE-SP) and is also responsible for the implementation and coordination of the Spatial Data Infrastructure Program for the State of São Paulo (IDE-SP). Emplasa has its own spatial data infrastructure, IDE-Emplasa, with unrestricted access, where its various geo systems are housed.

The Management of Cartography and Geomatics are the sectors of Emplasa that take care of the maintenance and administration of the cartographic collection, including the technical specification and the contracting of new surveys; this includes the acquisition of aerial and satellite imagery.

The Government of the State of São Paulo, through Emplasa, is responsible for the implementation, coordination, administration and execution of the actions related to the Cartographic System of the State of São Paulo (SCE-SP). In the context of these responsibilities, the tasks of Emplasa occasionally involve the acquisition of RS data. The Management of Geomatics manages, among other activities, the acquisition of satellite imagery aimed mainly at supporting specific projects. This satellite imagery is available internally through a geospatial database.

There is a general orientation in the Government of the State of São Paulo (Resolution Casa Civil n° 11 of 2005) so that the state body that purchases cartographic products shares such data with other departments and public institutions. It should be noted that Decree No. 61,486 of 2015 establishes that the agencies should consult Emplasa at the stage of elaboration of projects that require the production of geospatial data, to ensure standards are complied with and approved, and to eliminate duplication of effort and of resources. However, the acquisition in the scope of the State of São Paulo is not exclusive to Emplasa.

Depending on the demand, Emplasa acquires products or contracts private companies to perform the production of geospatial data. In other cases, the cartographic products are developed internally, using the technical knowledge inhouse. RS data and resulting products used and provided by Emplasa are as follows:

- Aerial imagery: used both internally in the development of projects and provided to the general public for legal proof, among other purposes,
- Satellite imagery: some specific internal projects require the acquisition of satellite imagery.

Emplasa has the Geospatial Data Sharing System (www.scd.emplasageo.sp.gov.br), which allows others to consult and acquire the cartographic products of its collection. Additionally, the provided spatial data can be visualized. There is also a so-called Public Service Sector, where it is possible to personally consult and purchase all the cartographic products of the Emplasa collection in analog and digital formats. Orthophotos and DEMs are not the only products derived from RS that are used by the public administration. The bodies make extensive use of aerial photographs, as well as systematic mapping (Cartographic Bases) and thematic maps (e.g. so-called Occupancy Mapping, land use mapping, and watershed protection area mapping) produced from satellite imagery.

In the scope of Decree nº 61.486 of 2015, Emplasa is responsible for the implementation of the São Paulo State Space Data Infrastructure—IDE-SP, where information about the data produced by the different agencies and entities of the State of São Paulo is cataloged and standardized, allowing direct access by the citizens (www.idesp.sp.gov.br) and the sharing of products through web-based services (WMS).

RS experts of Emplasa state there is no centralization of cartographic products in the collection of a single governing body. Some state and municipal agencies have cartographic collections, which they acquired over the years, to meet their internal demands. The experts explicate that there are ambitious efforts within the Government of the State of São Paulo to promote the centralization of cartographic production and the sharing of data between the different institutions to avoid duplicity in efforts to acquire products.

In relation to the municipality of São Paulo, RS experts of Emplasa state that the Municipal Secretariats of the Green and Environment, Urban Planning, Finance, Transportation and Works use the cartographic products of Emplasa as an input for their activities, mainly orthophotos and Systematic Mappings. The most commonly used products are systematic mapping products, in scales of greater detail, when available. The other municipalities (out of a total of 645 that make up the State of São Paulo) also use the various RS derived products provided by Emplasa.

Many municipalities and their respective departments do not use cartographic products in a scale of large detail (1: 1 000 and 1: 2 000); however, the scales of the products available in Emplasa can support their work as follows:

- Among other activities, the municipal environmental secretaries, as well as the State Secretariat for the Environment, request products for forest inventory, for inspection and licensing of works and occupation in areas of environmental preservation,
- The Municipal Work Secretaries use RS data and resulting products to promote, for example, the location of areas for the implementation of projects,
- The Municipal Planning Secretariats of the municipalities mainly seek knowledge of the territory for the implementation of public policies, and
- The Finance/Treasury Departments of the municipalities use RS data to support the updating of the Real Estate Registry (which allows the collection of Territorial Tax).

Emplasa records all assignments of cartographic products. In the assignment of digital files, both for public administration bodies and for teachers, students and ordinary citizens, there is a record of the justification of the request, in which the applicants need to describe the intended use for the requested datasets.

Emplasa receives requests for RS data and resulting products from different users: federal, state and municipal public administration bodies, teachers, students, ordinary citizens and private entities.

Emplasa can provide free cartographic products to clients ranging from institutions and entities of the federal state, the public municipal administration, to teachers and students. For individuals (individual and private entities), the products are sold according to an established price list. In the Spatial Data Infrastructure of the State of São Paulo—IDE-SP (www.idesp.sp.gov.br) the user obtains direct access to the products through web-based services (WMS).

Emplasa has the Geospatial Data Sharing System (www.scd.emplasageo.sp.gov.br), which allows the consultation and acquisition of the RS products and the visualization of the spatial data produced in the projects developed in-house. There is also a public service sector where it is possible to personally consult and purchase all available cartographic products, in analog and digital formats.

In the scope of State Decree nº 61.486 of 2015, Emplasa is responsible for the implementation of the São Paulo State Space Data Infrastructure—IDE-SP, where RS data and resulting products, produced by the different agencies and entities of the state of São Paulo, are cataloged and made accessible to the citizens (www.idesp.sp.gov.br).

RS experts of the state of São Paulo explicate that RS data have always been important for the public sector, although the implementation of the respective instruments, software, and techniques is very expensive. The RS experts state that sharing and accessing information has become easier, and new technologies have been incorporated to increase the number of users of the produced data.

Emplasa, initially responsible for the Metropolitan Cartographic System, has an obligation to produce accurate geospatial data since 2015. One of the responsibilities is the production of an official space reference within the scope of the state of São Paulo. For this reason, the production of the Emplasa cartographic base cannot be supported or referenced in non-official data (such as those provided by Google

Street View, Google Maps and Bing Maps and others). However, Emplasa uses some of the mentioned services to support specific projects, such as the geocoding of addresses.

In the prefecture of São Paulo (Table 1.37), RS techniques are used in various types of activities: for planning in general or to support public policy execution, as well as for (estate) monitoring and surveillance activities in general. In particular, the most important areas dealing with this type of product are cartography, geo-processing (GIS) and tax. Within the prefecture of São Paulo (Portuguese: Prefeitura de São Paulo), there is no department exclusively responsible for the acquisition of RS data and resulting products. Instead, each department makes this purchase according to its need. Thus, the acquisition depends on the specific work or project. Several departments (Portuguese: Secretarias) use and purchase aerial photos, such as the Department of the Environment, the Finance Department, and the inspection sector. The most important RS data are aerial photographs because of their spatial resolution. In 2017, the Finance Department and the Department of Green and the Environment acquired LiDAR data along with aerial photographs. Data management and distribution are organized in that images purchased by any department are generally distributed to any interested parties that would like to use them. The GeoSampa platform also distributes some aerial images both via an intranet and by making them publicly available through the internet, thus, accessible to any interested citizen. There are several products derived from RS data in use. Not all are produced by a particular department. Among others, RS data are usually used in a geoprocessing environment (GIS) to support public policy planning and monitoring.

Table 1.37 Overview of the prefecture of São Paulo

Overview	
Continent	South America
Country	Brazil
National GDP per capita in USD	15,484
National population	190,700,000
City population	12,180,000
Institution/City	Prefecture of São Paulo
Department responsible for remote sensing data and resulting products	The Geosampa Platform (maintained by the Municipal Department of Urbanism and Licensing —SMUL)
Administrative area in km^2	1,521
Number of used remote sensing platforms	2
WebGis Service	Yes
Services like Google Street View, Google Maps or Bing Maps	Yes
Considering the importance of RS within public authorities as	Increasing

The GeoSampa platform is maintained by the Municipal Department of Urbanism and Licensing—SMUL. Geosampa is a portal that follows the Strategic Master Plan, gathering georeferenced data about the city of São Paulo, including about 12 thousand urban facilities, the public transport network, geotechnical maps, and important data about the population, such as demographic density and social vulnerability. The portal is coordinated by the Department of Production and Analysis of Information (DEINFO) and developed by the Information and Communication Technology Company of the Municipality of São Paulo (PRODAM) (Secretaria Municipal de Urbanismo e Licenciamento 2018).

Such datasets are distributed through the GeoSampa platform (http://geosampa.prefeitura.sp.gov.br) to all sectors of the prefecture of São Paulo. However, since no particular department is responsible for the management of RS data and resulting products, these data can be passed on in other ways as well (e.g. personal requests).

In a nutshell, RS data and geodata in general are used for various activities, such as planning activities of various departments, for inspection activities, environmental monitoring, mapping of the road system, control of bus lines, attendance of traffic occurrences, etc.; there is a great diversity of themes.

There is no documentation regarding which city departments/divisions use RS data nor for which tasks or activities they are used. However, RS experts of the prefecture of São Paulo are certain that such data is widely disseminated and used. In addition, a working group on Geographic Information Systems of the Municipality of São Paulo was created in order to bring together the experiences and products that are gained and elaborated at the different departments.

Currently, most of the products purchased are available for use and download on the GeoSampa platform. Because there is both an intranet and an internet version, printed form requests are obsolete, and the data distribution to private entities, such as companies, extra-faculty (non-university) research institutes, university institutes, or individuals is not documented, as such, anymore. In general, the largest users are researchers, as well as companies that work with spatial data (GIS). In technological terms, the prefecture uses Web Map Services and Web Feature Services (WMS/WMF) for the sake of dissemination, following the standards of the Open Geospatial Consortium (OGC).

The data are available free of charge, except for those that may be subject to copyright. In such cases, an agreement can be established for use only, which is always free of charge, though.

RS experts of the prefecture of São Paulo consider the need for and the use of spatial information as increasing; especially, with the popularization of services such as Google Maps and the use of navigation/routing software, such as Here, Waze and Google Maps. The RS experts note that the acquisition of geoprocessing software was a barrier in the past, due to its high cost, which was overcome with the emergence of QGIS. It is noted that the popularization of these services has led to an increasing use of products derived from RS.

1.4.3.3 Ecuador

The following information was provided by RS experts of the Directory of Policies and Environmental Planning (Spanish: Dirección de Politicas y Planeamiento Ambiental) of the Department of Environment of the Quito Mayor's office. The municipality of Quito (Table 1.38) uses information from remote sensors for the multitemporal analysis of urban growth, deforestation, agricultural frontier growth, the loss of Paramo (alpine meadows), and wildfires. These operations are done in the unit of Knowledge Management. The analyses developed using RS are used for the planning of land use and land cover. RS is also used for the definition of mitigation projects or adaptation caused by environmental and climatic impacts; these can be risk control, reforestation or environmental remediation. This information is then used by other sectors such as the Department of Habitat and Housing Territory; Department of Security and Governability, Department of Planning, and the Metropolitan Institute for Urban Planning. Generally, the Department of Habitat and Housing Territory produces spatial information products on a scale of 1:1000 (urban scope), the Department of Mobility provides data on a scale of 1:100 (detail scope), and the Department for Environment provides data on a scale of 1:25.000 (rural scope). The implemented RS data generally consist of medium and high-resolution images and aerial images restored and generated to 3D models. Data management and distribution are organized by means of an "Open Data" platform. However, data are also provided by means of personal requests. Coordination meetings are organized to solve territorial problems between colleagues from different sectors. These meetings are considered to be more effective than the anonymous provision of datasets by means of online platforms, since there

Table 1.38 Overview of the city of Quito

Overview	
Continent	South America
Country	Ecuador
National GDP per capita in USD	11,617
National population	14,483,000
City population	2,239,000
Institution/City	Quito
Department responsible for remote sensing data and resulting products	Directory of Policies and Environmental Planning
Administrative area in km^2	4,183 (Distrito Metropolitano de Quito)
Number of used remote sensing platforms	3
WebGis Service	Yes
Services like Google Street View, Google Maps or Bing Maps	Yes
Considering the importance of RS within public authorities as	Remaining steady

Table 1.39 Tasks and activities RS data are used for

Tasks and activities
Monitoring of fires
Monitoring deforestation
Advance of the urban frontier—monitoring of urban growth
Estimation of the impact of greenhouse gases caused by the change of land use
Determination of ecological corridors
Control of compliance with environmental regulations

is still no common working culture within all institutions and departments regarding the search for spatial information on the web. Satellite imagery and UAV imagery are used for forest management and for the control of stone construction material utilization. RS derived products refer to the satellite imagery as such, the DEM, and the DSM; no other types of RS sensors are used.

The Directory of National heritage, the Directory of Climate Change and the Directory of Coastal and Marine Coordination use the service of the Directory of Policies and Environmental Planning; e.g. for indicators that allow territorial dynamics to be monitored. Other activities RS data are used for are as follows (Table 1.39).

Beside the open data platform (http://gobiernoabierto.quito.gob.ec/), there is an environmental catalog with updated information for internal use in the Department of Environment. Currently, a GIS web platform is being developed that will allow interaction between internal users and citizenship in the future. Public entities use the services of the Directory of Policies and Environmental Planning. In addition, students from universities apply for RS data. Personal requests can be made using the following website: http://www.quitoambiente.gob.ec/ambiente/index.php/politicas-y-planeacion-ambiental/atlas. RS experts of the city of Quito consider RS data as a key requirement for monitoring and responding to territory planning and regard the future importance of RS within public authorities as stable. Services like Google Street View, Google Maps or Bing Maps are used on a daily basis within the municipal organization.

1.4.3.4 Uruguay

In the City of Montevideo (Table 1.40), the Geomatics Service (Spanish: Servicio de Geomática) plays the main role in the acquisition of RS data and resulting products. The Geomatics Service is a unit of the Spatial Planning Division (Spanish: División Planificación Territorial) of the city administration (Spanish: Intendencia) of Montevideo. Due to institutional policy, all information has to be made available both to the internal departments of the city administration and to external users. The RS experts of Montevideo explicate that this policy makes it necessary to define standards and to ensure the interoperability of the information of image layers with other datasets of information that are (already) published.

Table 1.40 Overview of the city of Montevideo

Overview	
Continent	South America
Country	Uruguay
National GDP per capita in USD	22,563
National population	3,440,157
City population	1,300,000
Institution/City	Montevideo
Department responsible for remote sensing data and resulting products	Geomatics Service (on behalf of the Spatial Planning Division)
Administrative area in km^2	201 (Departamento de Montevideo: 500)
Number of used remote sensing platforms	3
WebGis Service	Yes
Services like Google Street View, Google Maps or Bing Maps	Yes
Considering the importance of RS within public authorities as	Increasing

In general, other municipal divisions such as Sanitation, Environmental Development and others do not autonomously sign-off contracts for the acquisition of RS data or resulting products. In any case, other municipal departments and divisions coordinate their activities in this field with the Geomatics Service (GS). Contracts may refer to spatial information (e.g. point data) for a specific project that is conducted in coordination with the GS. However, in no case, other departments will acquire RS data or resulting products for the entire territory of Montevideo themselves. RS data and resulting products used and provided by the GS are as follows:

- Aerial imagery: the city administration of Montevideo has maintained aerial photogrammetric coverage in a frequency of at least every 5 years. These are aerial images are in a 23 × 23 cm format at 1: 10.000 scale,
- UAV data: drones have been incorporated to carry out occasional surveys for cartographic updates and/or inspections; the actual drone models are a Leica brand exacerbator, Aibotix model, aDJI quad model, and a quad model Phantom 4 Pro,
- Satellite imagery: acquisition of satellite imagery (e.g. planet data) is planned for automated change detection.

The GS provides more than 350 levels of geo-referenced information (e.g. DEM, DSM, orthophoto), including their respective metadata, through its website. The layers are produced by means of RS data such as aerial imagery and other datasets such as contour lines. Both the specific demand of some users is addressed, and the photogrammetric restitution of constructions is carried out by traditional methods using the LPS software (Leica Photogrammetry Suite). Recently, the GS started to produce orthophotos, digital terrain models and digital elevation models in-house by means of the acquisition of UAV data and the respective processing software. The following departments of the city of Montevideo use the services of the GS and, thus, RS data and resulting products, respectively (Table 1.41).

Besides this, the subordinate municipalities (third level of government) and the Zonal Communal Center Services use the services of GS for the daily management of their respective territories to varying degrees. RS data and resulting products are used in the municipality for the development of the Territorial Ordinance Plan for urban planning and territorial planning, by the Land and Habitat Division for the regularization of irregular settlements and by the Division of Transportation and Service of Public Spaces and Buildings for the realization of technical pre-project works.

RS experts of Montevideo emphasize that the access to all spatial information is free and the RS data and resulting products are free of costs for everybody. The acquisition process does not require any registration nor demands a statement of motivation for the use of provided spatial datasets. Besides the experts' knowledge about which municipal departments and divisions use RS data and resulting products, and due to the mentioned open access policy, there is no specific documentation regarding which city departments/divisions use RS data or for which tasks or activities they are used.

Requests for RS data and resulting products come from public institutions, other public users and private institutions. Requests from private entities occur as follows: requests from private companies linked to the delivery of goods to establish and improve their routings; requests of providers of cartography products for GNSS navigators, requests of educational institutions to carry out historical investigations based on aerial imagery (for example, the search for clandestine burial places of the time of the military dictatorship in Uruguay; this work is carried out by the anthropology chair of the University of Montevideo and uses historical photos of eventual burial sites to detect signs). RS data and resulting products are provided by means of two web portals; directly downloadable from sig.montevideo.gub.uy, and accessible through http://geoweb.montevideo.gub.uy/geonetwork/srv/es/main.home.

RS experts of Montevideo consider the importance of RS within public authorities as increasing. They underpin this with the fact that the technology regarding RS techniques and products has rapidly developed throughout the last decades and has become more important in general.

Table 1.41 Top departments and divisions that use RS data and resulting products

Departments and divisions
Spatial Planning Division
Land and Habitat Division
Division of Sanitation
Division of Transportation and Service of Public Spaces and Buildings

Services like Google Street View and Google Maps and Bing Maps are widely used services for consultation by different areas of the municipality. RS experts of Montevideo further note that the institutional viewer of the city administration incorporates a significant number of layers to be displayed on the Google Earth platform in dynamic .kml and .kmz formats.

1.4.4 Asia

1.4.4.1 Taiwan

In the city of Taipei (Table 1.42), the Department of Urban Development (DUD) of the Taipei City Government plays the main role in the acquisition of RS data such as aerial imagery and terrestrial data. The DUD has established a regulation of providing RS data and resulting products such as aerial imagery, topographic maps and DTMs. RS data and resulting products used and provided by the DUD are as follows:

- Aerial imagery: aerial imagery used since 1991, most recent acquisition years: 2013 and 2015, 2017 orthophoto,
- LiDAR data: DEM, DSM,
- Terrestrial data: Topographic maps.

Data management and the distribution of RS data are organized by means of both personal requests and web platform (http://www.historygis.udd.gov.taipei/

Table 1.42 Overview of the city of Taipei

Overview	
Continent	Asia
Country	Taiwan
National GDP per capita in USD	49,827
National population	23,268,000
City population	2,700,000
Institution/City	Taipei City
Department responsible for remote sensing data and resulting products	Department of Urban Development
Administrative area in km^2	272
Number of used remote sensing platforms	3
WebGis Service	Yes
Services like Google Street View, Google Maps or Bing Maps	Yes
Considering the importance of RS within public authorities as	Increasing

urban_e/map/) including a WMTS service. The RS experts of the city of Taipei state that departments usually apply for topographic maps and aerial imagery in the context of civil and urban engineering. Amongst others, the following departments usually use the service of the DUD: The Public Works Department, the Department of Transportation, and the Department of Information Technology. According to the RS experts of the city of Taipei, there is documentation regarding which city departments use RS data because the requests for the RS data and resulting products are recorded. However, no detailed information is available because there is no explicit documentation regarding which tasks such data are used for. Generally, the departments mentioned beforehand use RS data and resulting products for tasks in urban planning, civil engineering, and transportation planning.

Requests for remote sensing data come from communal areas (within public authorities) and from private entities. Besides the WMTS services offered by the DUD, the provision of RS data and resulting products are associated with costs.

RS experts of Taipei consider the need for and the use of spatial information as increasing. They underpin this opinion with the increasing demand for aerial imagery and topographic maps. Services like Google Street View are used within the DUD. Other departments of the Taipei city government also use Google Maps as base maps.

1.4.4.2 United Arab Emirates

In the city of Dubai (Table 1.43), the Dubai Municipality-GIS Department (DMGD) plays the main role in the acquisition of RS data and functions as an

Table 1.43 Overview of the city of Dubai

Overview	
Continent	Asia
Country	United Arab Emirates
National GDP per capita in USD	73,879
National population	9,577,000
City population	2,500,000
Institution/City	Dubai
Department responsible for remote sensing data and resulting products	Dubai Municipality-GIS Department
Administrative area in km^2	4114
Number of used remote sensing platforms	3
WebGis Service	No
Services like Google Street View, Google Maps or Bing Maps	Yes
Considering the importance of RS within public authorities as	Increasing (slightly)

Table 1.44 Tasks and activities RS data are used for

Tasks and activities
The Environmental departments used products for nature protection purposes
Tree cadaster or greenery products are used by the Horticulture department
The Planning and Building department usually uses products of high-resolution imagery

interface for other departments. RS data and resulting products used and provided by the DUD are as follows:

- Aerial imagery,
- 3D laser scan data (currently only bathymetric),
- High-resolution satellite imagery.

Data management and distribution is currently organized through personal request. However, there is an undergoing refinement towards a web portal. Besides the Orthophoto and the DEM/DSM, other types of vector data are also collected from high-resolution satellite imagery. Further, some RS projects were conducted regarding Geology and Hydrology. Products such as city heat maps, public green space and NDVI maps, urban climate maps, etc. are provided. Besides this, different (semi-) government departments in dependence of particular projects, e.g. the Roads and Transport Authority (RTA) or the semi-government company Nakheel (property developer) use the services of the GIS Department (Table 1.44).

There is no documentation regarding which city departments/divisions use RS data nor for which tasks or activities they are used. Request for RS data and resulting products mostly come from public entities and rarely from private entities. RS data are free of charge for government departments and for semi-government institutions; they are not priced yet for the private sector, which means only products that are free of charge can be provided to the public, which excludes satellite images. RS experts of the City of Dubai consider the importance of RS within public authorities as increasing slightly. Services like Google Street View, Google Maps or Bing Maps are used within other municipal departments.

1.4.5 Africa

1.4.5.1 South Africa

Most departments of the City of Cape Town (Table 1.45) make use of RS data, which is maintained/analyzed within their respective GIS sections or by professional staff (town planners, road engineers, electrical engineers, etc). The City of Cape Town operates an internal CityMap GIS viewer, which currently contains 228 spatial data sets, available to all city employees. The Information and Knowledge

Table 1.45 Overview of the city of Cape Town

Overview	
Continent	Africa
Country	South Africa
National GDP per capita in USD	13,498
National population	57,581,000
City population	3,740,000
Institution/City	Cape Town
Department responsible for remote sensing data and resulting products	The Information and Knowledge Management (IKM) Department (Geomatics Branch)
Administrative area in km^2	400
Number of used remote sensing platforms	3
WebGis Service	Yes
Services like Google Street View, Google Maps or Bing Maps	Yes
Considering the importance of RS within public authorities as	Increasing

Management (IKM) Department (Geomatics Branch) is responsible for the acquisition/ dissemination of RS data, such as aerial images, LiDAR data, etc. The IKM Department manages the geospatial tenders and functions as the interface to other departments. Other departments can make use of the tenders, for example, order high-density LiDAR data acquisition for powerlines. RS data and platforms used by the city administration of Cape Town are as follows:

- Annual aerial photography (8 cm GSD, optional processing of fourth NIR band),
- Oblique imagery (5 cm GSD) annually (Midas sensor),
- LiDAR data (3 years turn around cycle),
- UAV imagery.

Data management and distribution of RS data and resulting products are organized by means of corporate warehouse servers that hold all data. Internal access is provided via the CityMap viewer that contains 228 data sets. Desktop GIS users are able to connect to the server where they can access 365 datasets in total. Aerial photography is provided via the Apollo image server. Externally, datasets can be accessed via the Open Data Portal or by email request to the City Maps public counter. Besides the orthophoto, products such as DEM, DSM (Tin or Raster) and contours are derived from the LiDAR data. 3D building models of the city are captured photogrammetrically from stereo pairs (aerial photography). 2D and 3D building footprints are derived from the building models; height maps are derived from the LiDAR dataset. A change detection map is derived from the DSM (UltraMap software) using two datasets (e.g. 2016 vs. 2017 and 2017 vs. 2018).

Table 1.46 Tasks and activities RS data are used for

Tasks and activities
Planning of roads, infrastructure, powerlines, human settlements, bus routes, transport interchanges
Beach monitoring, waste site monitoring, disaster management, court evidence
Heritage assessments, building and land use management
Change detection (new) by Valuations Departments
NIR imagery (in future) by Nature Conservation and Environmental Management Departments

Almost all departments use the service and datasets of the IKM Department (Table 1.46).

Internal acquisition of RS is documented by means of an inventory list. Requests for RS data come from internal and external clients, as well as from other governmental institutes. Data can be acquired through the City of Cape Town Open Data Portal, which holds 66 spatial layers and 33 non-spatial layers. RS experts of the City of Cape Town consider the need for and the use of spatial information as increasing. Google Street View is used and integrated in the City Maps Oblique IDAN Viewer; additionally, open data portal data sets are available in . kmz, shapefile, xls and pdf formats. Here Maps are integrated with the external viewer.

1.4.6 Oceania

1.4.6.1 Australia

In the state of South Australia (Table 1.47), there is an inter-departmental committee responsible for improving collaboration between departments in terms of information and budget sharing for the acquisition of RS data and resulting products. The Department of Planning, Transport and Infrastructure (DPTI) mainly uses aerial imagery and resulting orthophotos. However, these data are coordinated by the Department for Environment and Water (DEW).

In Australia, the administrative levels are organized as follows: Federal—State—Local council. When it comes to urban planning and RS data acquisition, the capital of the state of South Australia, the City of Adelaide, is responsible for its own jurisdiction, until a project reaches a level of state significance, either through cost, scale or importance of the project. In this context, the City of Adelaide has an independent GIS department and collaborates with the DPTI in various projects.

Within the administration of South Australia, data are shared within government departments via a shared link to an image server. The RS experts of DPTI are not aware of any documentation regarding which departments use RS data and resulting products over the image server. DPTI does not produce any commercial products or

Table 1.47 Overview of the state of South Australia

Overview	
Continent	Oceania
Country	Australia
National GDP per capita in USD	47,047
National population	21,500,000
Federal population	1,700,000
Institution/City	State of South of Australia
Department responsible for remote sensing data and resulting products	The Department for Environment and Water (DEW)
Administrative area in km^2	983,500
Number of used remote sensing platforms	3
WebGis Service	Yes
Services like Google Street View, Google Maps or Bing Maps	Yes
Considering the importance of RS within public authorities as	Increasing

services; thus, all work is done for internal use only. By means of the web platform, which is hosted by DEW, RS data and products are offered to the public. Mapland (https://www.environment.sa.gov.au/Science/mapland) is the commercial outlet for DEW. It provides access to a range of printed products (topographic maps, map books, etc) and digital data, such as aerial photography, digital topographic data including contours, environmental and location information, a customized mapping service, GIS data (vector and raster digital map data), digital elevation data, satellite image data, roads and property information, and publications. RS experts of the DPTI consider the need for and the use of spatial information as increasing. Services like Google Street View, Google Maps or Bing Maps are used within the DPTI for a quick and free inspection of a property.

DEW is considered the lead agency for the acquisition of aerial and high-resolution satellite imagery (DigitalGlobe, Airbus, DMC etc.), LiDAR, radar and medium resolution satellite imagery (i.e. Landsat, Sentinel) for the South Australia (SA) Government and project partners. As mentioned before, the DEW manages a panel of preferred suppliers that have been pre-approved to provide spatial imagery services, which include: aerial/satellite imagery acquisition, LiDAR/hyperspectral services, data processing and analysis, emergency acquisition, delivery services and aerial film scanning. Other SA government departments can partner with the DEW to access the panel to acquire services. The individual SA Government agencies can procure similar services without going through the respective DEW panel. Nevertheless, the DEW is generally the primary point of contact for these reasons, and the DEW has a long history of acquiring imagery for the SA Government (DEW captured aerial photography over South Australia from

1949 to 2004 using its own aircraft and analog camera). RS data and resulting products used and provided by DEW are as follows:

- Aerial imagery: both aircraft and remotely piloted aircraft systems (RPAS), orthorectified imagery mosaics,
- Satellite imagery,
- LiDAR.

For imagery projects, orthorectified imagery mosaics are the most common deliverable, but normally these are dependent on the project. Raw multispectral data is used for environmental analysis.

Data management and the distribution of RS data is internally organized by means of a GIS and related databases, for example, orthomosaics can be delivered by a variety of means:

1. An internal subscription service available to all SA Government agencies called "ImageMapSA",
2. Web-applications, which can either be restricted to SA Government access only due to the sensitivity of the information "EnvMaps" or available to the general public through "NatureMaps" (https://data.environment.sa.gov.au/NatureMaps/Pages/default.aspx),
3. Personal requests through our commercial outlet "Mapland" (https://www.environment.sa.gov.au/Science/mapland).

Orthorectified image mosaics are the most common and widely-utilized product within the SA government. Precision elevation data from LiDAR or photogrammetric processing in the form of bare earth DEMs are most commonly used for tasks in urban development, urban planning and flood modeling. The SA Government also uses derived landcover products from Landsat and Sentinel satellites, as well as products developed from the Landsat Data Cube (or Digital Earth Australia: http://www.ga.gov.au/dea). The landcover change layers are also available through "EnvMaps" for those users without a respective desktop GIS software.

Other products are provided by DEW and some other agencies. The tasks and activities they are used for are as follows (Table 1.48).

Table 1.48 Tasks and activities RS data are used for

Tasks and activities
Benthic mapping using aerial (includes hyperspectral) and satellite imagery
Regional score cards assessing native vegetation/environmental health
Modeling LiDAR to assess coastal flooding due to climate change
Fauna counts
Urban heat mapping using thermal aerial imagery
3D building models to support urban development and planning

Table 1.49 Departments and divisions that use RS data and resulting products

Departments and divisions
Department for Environment & Water (DEW)
Department of Planning, Transport and Infrastructure (DPTI)
SA Water
SA Police
Renewal SA
Environment Protection Authority
Vinehealth Australia
Local government

The main stakeholders using RS data and resulting products are as follows (Table 1.49).

However, RS data and resulting products are also used by the wider government community, as the majority of the data procured has intellectual property to the SA Government or a license for the SA Government use where intellectual property cannot be given.

Documentation about which departments use RS data and for which tasks exists in a non-centralized way, as the procurement in the SA Government is project-based to meet a particular business requirement. However, data are not restricted to these projects as the intent is to procure remotely sensed data that is "purchased once, used many times".

Most of the requests for RS data and resulting products are internal requests and come from the SA Government agencies. Besides this, there are collaborations between universities and private companies.

Generally, image data is not free of charge. It is expected that stakeholders contribute budget to procure the data. If enough budget is not available, the acquisition does not proceed. Once data has been procured, it is shared within the SA government and project partners. It can also be accessed through a paid subscription, or under a data license agreement.

There are several web portals where RS data and resulting products can be acquired from Mapland (https://www.environment.sa.gov.au/Science/mapland), EnvMaps, NatureMaps and the Location SA Map Viewer can be used to view imagery, but imagery cannot be downloaded from these portals.

RS experts of the DEW consider the importance of RS within public authorities as generally increasing. They state there is more interest on how to access remotely sensed data. They explicate that transitional procurements methods are still most commonly used, but new subscription services are always being investigated from a whole government perspective. This includes subscription services from aerial companies in Australia, satellite imagery services from the likes of DigitalGlobe, Planet, Earth-I, etc. Services like Google Street View, Google Maps or Bing Maps are used.

1.5 Reasoning

Almost all city administrations globally use RS data and resulting products for tasks in urban planning on an operational basis. However, the organization of institutions responsible for the acquisition and distribution of such data strongly varies. Data management and the distribution to city departments often relates to the organizational structure of a city and to its financial capacities. In a global context, there is no clear tendency that larger cities or cities with more population necessarily use RS data more often or are more likely to host a particular land survey or GIS department that is exclusively responsible for RS data and geodata in general. It is worth mentioning that the information about a city, its departments, and their respective functions provided through their websites or other information material strongly differs from one country to another and has no relation to the size of a city or its potential financial or technical background. A summary of the findings is given in Table 1.50.

Aerial imagery is the most important RS data source for city administrations. The acquisition and interpretation of such data have a fixed place in most city administrations worldwide.

Satellite imagery is generally less frequently used than aerial imagery. RS experts often argued the geometric resolution is too low for urban planning applications, mainly in dense urban areas. In other cities, satellite imagery is used to

Table 1.50 Summary of the findings

Overview			
Object of investigation	Frequency		
Institutions/City departments contacted	73		
Institutions/City departments with comprehensive return of information	31 (return rate of 42%)		
Countries covered in the study	22		
Used remote sensing platforms	Aerial imagery		29
	Satellite imagery		17
	LiDAR		20
	Radar		1
	Hyperspectral		3
	Terrestrial laser scanning		10
	UAV		5
Services like Google Street View, Google Maps or Bing Maps are	None	Particular	All mentioned
	1	11	17
Considering the importance of RS within public authorities as	Decreasing	Steady	Increasing
	0	2	27 (29)
WebGIS products	Yes	No	
	28	1	

support aerial imagery, for example, to complement or substitute the acquisition of aerial imagery, perhaps between two respective acquisition dates. For this purpose, often the mere visual interpretation of the satellite imagery by RS experts is important. In addition to this, RS experts explicated that the costs for satellite imagery are still high when compared to aerial imagery; in many cases, satellite imagery is not operationally used or acquired and merely requested in relation to certain projects. In other countries, satellite imagery is additionally acquired for the monitoring of rapidly developing urban areas.

Radar data are hardly used by any city administration. I trace this back to the higher level of complexity of these RS data when compared to aerial imagery or LiDAR data. Until now, the advantages of radar data such as the acquisition regardless of weather conditions or light conditions apparently cannot countervail the demand of expert knowledge for the corresponding data (pre-) processing and interpretation.

Terrestrial laser scan data are partly used in municipalities. These data are often used in the context of preserving sites of historic interest or for tunneling constructions.

In many municipalities, LiDAR data and resulting products are already an inherent part of the utilized datasets. LiDAR data and resulting products are used for visibility analysis, urban climate-related studies or energy planning related projects. The acquisition of LiDAR data is generally conducted in longer time lags when compared to the acquisition of aerial or satellite imagery. I trace this back to the expenses for the corresponding acquisition flights and data preparation.

Out of 41 contacted countries, information regarding the use of RS and resulting products were received from 22 countries. One may argue that the number of countries participating in the study is low compared to the overall number of contacted countries, cities and institutions. I address this circumstance in the following section, and I want to pre-address potentially missing information and obstacles that occurred while conducting this study.

1.5.1 Diverse Perception of RS

While conducting this study, it was not a seldom response by the contacted city department (GIS, urban planning, city development, land survey, open data etc.) that they do not use RS data, methods or instruments in any form. Since some of these departments already stated on their website that they use aerial imagery for their city web map service, I started to doubt particular disclosures. Other departments responded that they do not use RS data but only aerial imagery and orthophotos. The given examples shall highlight that there is apparently still a gap of knowledge and a certain challenge regarding the perception about RS, RS methods and RS instruments. As mentioned in the introduction section, urban RS involves all activities regarding the acquisition of information about the urban environment and physical objects within urban areas through recording, measuring and interpreting electromagnetic radiation derived from contactless sensors. This definition encompasses the acquisition of aerial imagery, aerial laser scan data, terrestrial laser scan data, street view imagery, radar data, ground-penetrating radar, thermal imagery, traffic remote sensing data etc.

There is no doubt about the remaining uncertainties regarding to what extent RS data and resulting products are used. However, this study sheds light on these uncertainties and shows what RS data are acquired, which departments they are used by, and for which tasks with the city administration they are used.

1.5.2 Remaining Cities and Institutions

The attentive reader may have noticed that there are several cities or institutions that are listed in Table A1 but are not mentioned in the results section. From certain cities and institutions, no answer was received. Other cities and institutions answered to my request as follows:

- France, Paris: The Directory of Systems and Information Technology (DSTI) responded that the DSTI is responsible for the management and implementation of the City Geographic Information System, the management of the geographic data repository, and providing support services to the directorates of geographic information (data processing in France, spatial analysis, digital cartographies and web-applications intranet or internet). The exploitation of tools to produce business data or business studies is supported by the city's business divisions. At this stage, the DSTI does not use remote sensing as part of its missions.
- Estonia, Tallinn: The respective unit of the Tallinn Urban Planning Department uses aerial imagery and orthophotos to update the base map of Tallinn. An orthophoto map service is used as background information in different map applications by the city of Tallinn.
- Japan: The Geoinformation Authority of Japan (GSI) responded to my request by stating that the GSI is using RS for conducting research into understanding disaster situations and topographic and crustal deformation processes. However, the GSI does not use RS for urban planning or energy planning in the city administration. The GSI recommended I contact two other institutions, yet, I received no answers from either of them. According to a person in charge of the city of Kobe, RS is not used within the Kobe city administration.

One Welsh municipality responded that they "do not have the resources to assist anyone with their personal studies and they must use their limited resources to assist their respective citizens".

Although positive answers from other cities or institutions were received, the transmitted questionnaire was not returned to the author.

1.5.3 Sensitization and Future Use of RS Data

This study shall serve as a contribution to promote the use of RS among the public. Other ways to strengthen public perception in general, and to promote RS for city administrations in particular, are as follows:

Promote private companies and their establishment specializing in remote sensing services and related products,

Offer more courses related to RS in the following urban planning related careers: architecture, urban planning, regional planning, spatial planning, landscape planning, civil engineering; offer workshops that allow for training in handling and preprocessing of RS data accompanied by practical exercises.

Develop new tools to respond to the needs of urban planners for their daily tasks.

The latter is a crucial vehicle to promote the use of RS data and resulting products in municipalities. The resolution, the availability, the cost, the processing time, the field of application; these are all factors that researchers should pay specific attention to while developing new methods and tools in the future.

Such developments may be supported by the recognizable tendency of research in recent years that veers toward cloud computing in geospatial image processing and new approaches for earth observation data management. In this way, such research will support the ability of municipal departments to become independent from limitations of particular cost-intensive hardware and software.

This study shows that the knowledge and awareness for RS and its diverse applications already exists in many municipalities to a certain degree. Regarding the use of freely available satellite sensed data it can be stated that the use of these data often remains limited to certain areas of application of the respective departments. One reason may be found in the resolution of such data (Sentinel program, Landsat program). Thus, research and private companies are required to either produce suitable products with these kinds of data or to develop methods to allow for combining them with in-house data (land registers, air emission values, traffic low analyzes etc.) and hence develop new useful products that can support urban planners in operational areas.

Conclusion

I described the use of RS data and resulting products within city administrations on a global scale. RS data are used for urban planning applications worldwide to a certain extent. However, the organization of how RS data are used and the distribution of such data occasionally differs from one country to another, even from one city to another within the same country. The study revealed that the size of a city or the population size are not dependent variables when it comes to the organization of city administrations or the institutional areas of responsibility. This study provided a comprehensive overview of how city administrations use RS data and resulting products, and for which tasks these data are used. It is worth mentioning that it was not the objective of this study to present showpieces regarding perfect data management or institutional organizations. The administrative organization of a city may have been shaped politically, culturally or by a compound of both these factors. Thus, one organizational structure of one country does not necessarily represent a showpiece for all other countries. I conclude that the organizational structure itself is not the dependent variable when it comes to reasonable acquisition of RS data and reasonable efficient intercommunal data management. The (ideally regulated) exchange of information regarding the acquisition of RS data and resulting products is fundamental though. I consider it crucial that RS data and resulting products such as base maps, orthophotos, or other datasets, are shared in an intercommunal way by means of one central GIS database or web platform solution. The provision of datasets at one central platform anticipates unnecessary data acquisitions and facilitates the discovery and use of existing datasets and layers. Therefore, the exchange of data and information has to be assured within a city administration or within a respective governmental entity.

RS data and resulting products are used as follows:

- impervious surface mapping,
- water and wastewater management,
- land cover analysis products,
- land cover mapping,

A. Lehner, *The Operational Use of Remote Sensing in Municipalities*, SpringerBriefs in Geography, https://doi.org/10.1007/978-3-030-23800-1

- flood protection (calculation of discharges),
- green area planning, urban climate analysis,
- control of compliance with building regulations,
- control of compliance with environmental regulations,
- monitoring of fires,
- monitoring deforestation,
- determination of ecological corridors,
- planning of roads,
- planning of infrastructure in general,
- planning of powerlines,
- planning of human settlements,
- planning of bus routes,
- planning of transport interchanges,
- beach monitoring,
- waste site monitoring,
- disaster management,
- court evidence,
- fauna counts,
- addressing,
- monitoring of the redevelopment of brownfield properties,
- assessment of storm water and ground water behavior,
- noise cadaster,
- vegetation monitoring,
- green roof potential cadaster,
- tree cadaster,
- spatial planning in general,
- virtual site visits,
- aerial imagery for ground-truthing of new spatial datasets,
- production thematic maps,
- green space monitoring,
- determination of drainage basin sizes and drainage basin flow directions,
- heritage assessments,
- advance of the urban frontier—monitoring of urban growth,
- estimation of the impact of greenhouse gases caused by the change of land use,
- building and land use management in general,
- change detection, and
- benthic mapping.

Most of the RS experts stated the importance of RS data and resulting products will increase within public authorities in the future. They motivate this with the increasing demand for aerial imagery and topographic maps. The public awareness for such data and methods appear to increase and, in fact, more governmental reports demand the documentation of areas and locations by means of imagery or maps. None of the RS experts stated a decrease of the importance of RS data and resulting products. Those experts that indicated the importance of such data as remaining steady justified their response by explaining that there are still either gaps in departmental knowledge or less interest within other departments for RS data and resulting products. Other reasons were found in the already comprehensive provision of RS data and related information within the governmental entities.

During this study, I came across the diverse expert knowledge regarding what RS is, what methodologies are state-of-the-art and where RS contributes to the tasks of city administrations. Therefore, the identified gap in knowledge about what RS is, what it consists of, and how beneficial these data may be for a large variety of municipal departments, must be closed. Hence, this study contributes to raising such awareness and demonstrates that RS data and resulting products are more than just nice background pictures within a vector map.

Appendix

See Tables A1 and A2.

Table A1 Contacted cities and institutions for conducting the RS expert interviews

Continent	Country	City or institution contacted
Africa	Egypt	Cairo
	South Africa	Cape Town
	Kenya	Kinshasa
	Namibia	Windhoek
	Morocco	Marrakech
Asia	India	Indian Institute of Remote Sensing
	Indonesia	Jakarta
	Japan	Fukuoka
		Geospatial Information Authority of Japan
		Kobe
		Osaka
		Spatial Information Science Research Faculty, University of Tsukuba
	Singapore	Ministry of Communications and Information
	Sri Lanka	Sri Lanka Survey Department
	United Arabian Emirates (UAE)	Dubai
	Vietnam	Hanoi
		Ho Chi Minh City
	Taiwan	Taipei

(continued)

A. Lehner, *The Operational Use of Remote Sensing in Municipalities*, SpringerBriefs in Geography, https://doi.org/10.1007/978-3-030-23800-1

Table A1 (continued)

Continent	Country	City or institution contacted
Australia	Australia	Western Australian Land Information Authority
		Spatial Services division New South Wales
		ACT Government business unit (Canberra)
		Government of South Australia
		Brisbane
	New Zealand	Auckland
Europe	Austria	Vienna
		Graz
		Salzburg
	Belgium	Brussels
	Estonia	Tallinn
	Germany	Erfurt
		Munich
	France	Lyon
		Paris
		Strasbourg
		Regional Government Auvergne-Rhône-Alpes
	Hungary	Lechner Knowledge Center (Lechner Tudásközpont)
		Szent István University
		(National) Department of Geodesy, Remote Sensing and Land Offices (Földmérési, Távérzékelési és Földhivatali Főosztály)
	Iceland	Reykjavik
	Ireland	Dublin
	Italy	Milan
		Naples
	Norway	Oslo
	Poland	Institute of Spatial Management and Housing (Instytut Gospodarki Przestrzennej i Mieszkalnictwa)
		Office of Surveying and Cadastre (BGiK - Biuro Geodezji i Katastru)
	Portugal	Lisbon
	Romania	Bucharest
	Spain	Madrid
		Saragossa
	Sweden	Stockholm
	United Kingdom	Birmingham
		Cardiff
		Glasgow
		Ordnance Survey
		Barnsley Metropolitan Borough Council

(continued)

Table A1 (continued)

Continent	Country	City or institution contacted
North America	Canada	Calgary
		Toronto
		Vancouver
	Cuba	Instituto de Planificación Física, Havana
	Mexico	Mexico City
		GISMEXICO Geographic Solutions
	USA	Anchorage
		Denver
		Los Angeles County
		Miami Dade County
		City of Miami
South America	Argentina	Buenos Aires
	Bolivia	La Paz
	Brazil	Municipal Secretary of Urban Development (Secretária Municipal de Desenvolvimento Urbano de a Prefeitura Municipal de São Paulo)
		Cartography Unit—São Paulo State Government (EMPLASA - Empresa Paulista de Planejamento Metropolitano SA—Gerência de Cartografia)
	Ecuador	Quito
	Trinidad & Tobago	Port of Spain
	Uruguay	Montevideo

Table A2 Websites of contacted cities and institutions

Continent	Country	City or institution contacted	
Africa	South Africa	Cape Town	http://www.capetown.gov.za/Departments/Information%20and%20Knowledge%20Management%20Department
Asia	Taiwan	Taipei	https://english.udd.gov.taipei/
Australia	Australia	Government of South Australia	http://spatialwebapps.environment.sa.gov.au/naturemaps/?locale=en-us&viewer=naturemaps

(continued)

Table A2 (continued)

Continent	Country	City or institution contacted	
Europe	Austria	Graz	https://www.graz.at/cms/ziel/8044336/DE
		Salzburg	https://www.stadt-salzburg.at/internet/politik_verwaltung/bauen/stadtvermessung/vermessung_und_geoinformation_467673.htm
		Vienna	https://www.wien.gv.at/stadtentwicklung/stadtvermessung/
	Belgium	Brussels	https://cirb.brussels/
	France	Strasbourg	http://www.sig.strasbourg.eu/
	Germany	Erfurt	https://www.erfurt.de/ef/de/rathaus/sv/aemter/stelle-106.htmc
		Munich	https://www.muenchen.de/rathaus/Stadtverwaltung/Kommunalreferat/geodatenservice.html
	Hungary	Lechner Knowledge Center (Lechner Tudásközpont)	http://lechnerkozpont.hu/
		(National) Department of Geodesy, Remote Sensing and Land Offices (Földmérési, Távérzékelési és Földhivatali Főosztály)	http://www.ftf.bfkh.gov.hu/portal_en/
	Iceland	Reykjavik	https://borgarvefsja.reykjavik.is/borgarvefsja/
	Italy	Naples	http://www.comune.napoli.it/flex/cm/pages/ServeBLOB.php/L/IT/IDPagina/202
	Norway	Oslo	https://www.oslo.kommune.no/politikk-og-administrasjon/etater-og-foretak/plan-og-bygningsetaten/
	Poland	Office of Surveying and Cadastre (BGiK - Biuro Geodezji i Katastru)	http://www.mapa.um.warszawa.pl/BGIK/?utm_source=um.warszawa.pl&utm_medium=redirect%2B301&utm_campaign=ex%2Bhtaccess
	Portugal	Lisbon	http://geodados.cm-lisboa.pt/
	Spain	Madrid	http://www.comunidad.madrid/servicios/mapas/geoportal-comunidad-madrid
	United Kingdom	Barnsley Metropolitan Borough Council	https://www.barnsley.gov.uk/barnsley-maps/barnsley-interactive-maps/

(continued)

Table A2 (continued)

Continent	Country	City or institution contacted	
North America	Canada	Calgary	http://www.calgary.ca/CS/IIS/Pages/home.aspx
		Toronto	https://www.toronto.ca/city-government/data-research-maps/open-data/
	Mexico	Mexico City	http://www.paot.org.mx/
	USA	Denver	https://www.denvergov.org/content/denvergov/en/geographic-information-systems.html
		Los Angeles County	https://egis3.lacounty.gov/dataportal/lariac/
South America	Bolivia	La Paz	http://geo.gob.bo/portal/
	Brazil	Municipal Secretary of Urban Development (Secretária Municipal de Desenvolvimento Urbano de a Prefeitura Municipal de São Paulo)	http://geosampa.prefeitura.sp.gov.br/PaginasPublicas/_SBC.aspx
		Cartography Unit—São Paulo State Government (EMPLASA - Empresa Paulista de Planejamento Metropolitano SA—Gerência de Cartografia)	www.idesp.sp.gov.br
	Ecuador	Quito	http://www.quitoambiente.gob.ec/ambiente/index.php/politicas-y-planeacion-ambiental
	Uruguay	Montevideo	http://sig.montevideo.gub.uy/

Glossary

CIR Color Infrared Imagery is a false color photograph that shows the reflected electromagnetic waves from an object accordingly: Near Infrared (NIR), which is invisible to the human eye, as red; green light as blue; red light as green (Statewide Mapping Advisory Committee 2011).

DEM Digital Elevation Model, a 3D computer graphics representation of a terrain's surface.

DSM Digital Surface Model, a digital model representing the entire surface of an area that may include bare ground and elevated objects.

DTM Digital Terrain Model, a digital model covering the bare ground surface only.

GIS A geographical information system is a computer-based system that processes geographical information (MacEachren and Taylor 2013).

Geodatabase A database designed to store, query, and manipulate geographic information and spatial data (Peters 2008).

GSD Ground Sampling Distance, represents the distance in a digital photo between pixel centers measured on the ground; it is a measure of resolution limitations due to sampling (Leachtenauer and Driggers 2001).

LiDAR Light detection and ranging, a remote sensing method to measure distance to a target by illuminating the target with pulsed laser light and measuring the reflected pulses with a sensor. Differences in laser return times and wavelengths are used to generate digital 3D representations of the target.

LoD Level of detail, a measure to indicate the geometric detail of a (3D) model, primarily of buildings. It is used to define a series of different representations of real world objects, and to suggest how thoroughly they have been acquired and modelled (Biljecki et al. 2014).

A. Lehner, *The Operational Use of Remote Sensing in Municipalities*, SpringerBriefs in Geography, https://doi.org/10.1007/978-3-030-23800-1

nDSM Normalized Digital Surface Model, the result of the difference of the Surface and Terrain Models.

NDVI Normalized Difference Vegetation Index, indicates the existence of green vegetation and the density of green for a certain area.

NIR Near Infrared, an array of the electromagnetic wavelength spectrum usually ranging between 750 and 1400 nm.

RGB Red Green Blue, color information of remotely sensed imagery.

RS Remote Sensing: the acquisition of information about an object or phenomena without physically touching it.

UAV Unmanned Aerial Vehicle, commonly also known as drone.

WMS Web Map Service is a standard protocol developed by the Open Geospatial Consortium in 1999 for serving georeferenced map images over the internet (Scharl and Tochtermann 2009).

References

Agrawal D, Kettinger WJ and Zhang C (2014) The openness Challenge: why some cities take it on and others don't. Poster Presentation at 20th Americas Conference on Information Systems (AMCIS 2014), Savannah, Georgia, USA, August 7–9, 2014.https://pdfs.semanticscholar.org/e2fa/121be3a780b5b8a6338bb6c2cf3e7ac2e589.pdf

Baruth B, Royer A, Klisch A, Genovese G (2008) The use of remote sensing within the MARS crop yield monitoring system of the European Commission. Int Arch Photogramm Remote Sens Spat Inf Sci 37:935–940

Bhatta B (2010) Analysis of urban growth and sprawl from remote sensing data. Springer Science & Business Media, New York

Biljecki F, Ledoux H, Stoter J, Zhao J (2014) Formalisation of the level of detail in 3D city modelling. Comput Environ Urban Syst 48:1–15

Bogner A, Menz W (2009) The theory-generating expert interview: epistemological interest, forms of knowledge, interaction. Interviewing experts. Springer, New York, pp 43–80

CIRB (2018) CIRB-CIBG-BRIC. In: The bric. https://bric.brussels/en/the-bric

Comunidad de Madrid (2018) GEOPORTAL DE LA INFRAESTRUCTURA DE DATOS ESPACIALES DE LACOMUNIDAD DE MADRID. In: Geoportal idem. http://www.madrid.org/cartografia/idem/html/visores.htm

DiCicco-Bloom B, Crabtree BF (2006) The qualitative research interview. MEDU Med Educ 40:314–321

Esch T, Taubenböck H, Heldens W et al. (2010) Urban remote sensing–how can earth observation support the sustainable development of urban environments? In: Proceedings. pp 1–11

Franchino N (2018) LARIAC—Los Angeles Region Imagery Acquisition Consortium Program. In: Los Angeles County GIS Data Portal. https://egis3.lacounty.gov/dataportal/lariac/

Fund, Internal monetary. World economic outlook database. World Econ Finance Survey, (2018) https://www.imf.org/external/pubs/ft/weo/2018/02/weodata/index.aspx

GeoBolivia (2018). In: GeoBolivia. http://geo.gob.bo

Governance and legal status of the Ordnance Survey (2018). In: Governance. https://www.ordnancesurvey.co.uk/about/governance/index.html

Herold M, Scepan J, Clarke KC (2002) The use of remote sensing and landscape metrics to describe structures and changes in urban land uses. Environ Plann A 34:1443–1458

Jobst M (2013) Thermalbildauswertung Wien Verwendungsmöglichkeiten von Thermalbildern im städtischen Raum. Wien, Univ. für Bodenkultur, Masterarb., 2013. https://obv-at-ubbw.alma.exlibrisgroup.com/discovery/openurl?institution=43ACC_UBBW&rfr_id=info:sid%2Fsummon&rft_dat=ie%3D2141242780003345,language%3DEN&svc_dat=CTO&u.ignore_date_coverage=true&vid=43ACC_UBBW:Services

A. Lehner, *The Operational Use of Remote Sensing in Municipalities*, SpringerBriefs in Geography, https://doi.org/10.1007/978-3-030-23800-1

Kadhim N, Mourshed M, Bray M (2016) Advances in remote sensing applications for urban sustainability. Euro-Mediterranean J Environ Integr 1:7
Klisch A, Atzberger C (2016) Operational drought monitoring in Kenya using MODIS NDVI time series. Remote Sens 8:267
Landtag W (2001) Wiener Stadtentwicklungs-, Stadtplanungs-und Baugesetzbuch (Bauordnung für Wien–BO für Wien)
Leachtenauer JC, Driggers RG (2001) Surveillance and reconnaissance imaging systems: modeling and performance prediction. Artech House
Lehner A, Steinnocher K (2016) Urban remote sensing and energy planning—Doctoral Consortium Contribution. In Doctoral Consortium—DCGISTAM, (GISTAM 2016) ISBN, pp 8–11. https://www.scitepress.org/PublicationsDetail.aspx?ID=Y/a2Ce9CJlU=&t=1
Lune H, Berg BL (2017) Qualitative research methods for the social sciences (9th, Global Edition). *Essex: Pearson Education Ltd.* https://www.amazon.de/Qualitative-Research-Methods-Social-Sciences/dp/0205809383
MacEachren AM, Taylor DRF (2013) Visualization in modern cartography. Elsevier, Amsterdam
Martini BA, Cocks TD, Cocks PA, Pickles WL (2004) Operational airborne hyperspectral remote sensing for global geothermal exploration. In: Geoscience and remote sensing symposium, 2004. IGARSS'04. Proceedings. 2004 IEEE International. IEEE
Netzband M, Stefanov WL, Redman C (2007) Applied remote sensing for urban planning, governance and sustainability. Springer Science & Business Media, New York
Olsen RB, Bugden P, Andrade Y, et al. (1995) Operational use of RADARSAT SAR for marine monitoring and surveillance. In: Geoscience and remote sensing symposium, 1995. IGARSS'95.'Quantitative remote sensing for science and applications', International. IEEE, pp 224–226
Peters D (2008) Building a GIS: System architecture design strategies for managers. ESRI, Inc
Saitoh S-I, Mugo R, Radiarta IN et al (2011) Some operational uses of satellite remote sensing and marine GIS for sustainable fisheries and aquaculture. ICES J Mar Sci 68:687–695
Sakaiya E, Inque Y (2013) Operational use of remote sensing for harvest management of rice. J Remote Sens Soc Jpn 33:185–199
San-Miguel-Ayanz J, Ravail N (2005) Active fire detection for fire emergency management: potential and limitations for the operational use of remote sensing. Nat Hazards 35:361–376
Scharl A, Tochtermann K (2009) The geospatial web: how geobrowsers, social software and the Web 2.0 are shaping the network society. Springer Science & Business Media, New York
Secretaria Municipal de Urbanismo e Licenciamento (2018) GeoSampa: Noticias. In: Prefeitura de São Paulo - Secretaria Municipal de Urbanismo e Licenciamento. http://www.prefeitura.sp.gov.br/cidade/secretarias/urbanismo/noticias/?p=247194
Specter C (1988) Managing remote sensing technology transfer to developing countries: a survey of experts in the field. Photogrammetria 43:25–36
Stadtvermessung Wien—MA 41 (2016) Geodatenviewer der Stadtvermessung Wien. In: Stadtplan. https://www.wien.gv.at/ma41datenviewer/public/start.aspx
Statewide Mapping Advisory Committee WG for OP (2011) Using Color Infrared (CIR) Imagery —A guide for understanding, interpreting and benefiting from CIR imagery
Tooke TR, Coops NC (2013) A review of remote sensing for urban energy system management and planning. In: *Joint Urban Remote Sensing Event 2013*. IEEE, 2013. pp 167–170. https://ieeexplore.ieee.org/abstract/document/6550692, https://doi.org/10.1109/JURSE.2013.6550692
Wallace L, Lucieer A, Watson C, Turner D (2012) Development of a UAV-LiDAR system with application to forest inventory. Remote Sens 4:1519–1543
World Bank Group (2017) World development indicators 2017. World Bank Publications, Washington
WorldPopulationReview (2016) Total population by country 2018. In: Worldpopulationreview. http://worldpopulationreview.com/countries. Accessed 16 Oct 2018
Yang X (2011) Urban remote sensing: monitoring, synthesis and modeling in the urban environment. John Wiley & Sons, New York

MIX
Papier aus verantwortungsvollen Quellen
Paper from responsible sources
FSC® C105338

If you have any concerns about our products,
you can contact us on
ProductSafety@springernature.com

In case Publisher is established outside the EU,
the EU authorized representative is:
Springer Nature Customer Service Center GmbH
Europaplatz 3, 69115 Heidelberg, Germany

Printed by Libri Plureos GmbH
in Hamburg, Germany